ASTRONOMY IN MINUTES

GILES SPARROW

ASTRONOMY IN MINUTES

GILES SPARROW

Quercus

The constellation maps throughout this book use the symbols displayed below to identify stars of different magnitude and other celestial objects:

✸	Star ≤ mag 0.0	●	Star ≤ mag 4.0	◉	Nebula
✸	Star ≤ mag 1.0	•	Star ≤ mag 5.0	⬭	Open star cluster
★	Star ≤ mag 2.0	·	Star ≤ mag 6.0	⬭	Globular cluster
★	Star ≤ mag 3.0			▬	Galaxy

By convention, astronomers use the 24 letters of the Greek alphabet, listed below, to identify the brightest stars in a constellation (though the application of these rules is somewhat inconsistent and the present system owes a good deal to historical accident).

α - alpha	η - eta	ν - nu	τ - tau
β - beta	θ - theta	ξ - xi	υ - upsilon
γ - gamma	ι - iota	o - omicron	φ - phi
δ - delta	κ - kappa	π - pi	χ - chi
ε - epsilon	λ - lambda	ρ - rho	φ - psi
ζ - zeta	μ - mu	σ - sigma	ω - omega

CONTENTS

Introduction

Astronomy is the oldest of all the sciences, and yet it is also one of the most rapidly changing – the celestial objects in our night sky have been a source of fascination since before the dawn of recorded history, but our understanding of what these objects actually *are* – and the mechanisms that create, shape and sustain them alongside a host of shorter-lived phenomena – is in many cases only a few decades old, and inevitably subject to the change and development seen in all scientific theories. The result is a unique hybrid science – astronomy can mean many things to different people, from a repository of ancient star lore to a field of cutting-edge theoretical and technological development.

Picking just 200 individual entries to encompass astronomy as a whole is therefore something of a challenge – most books tackling the subject tend to fall on one side or another of the broad divide between practical observing handbooks and more descriptive guides to the Universe. Either approach on its own could happily fill this book's extent, but the stated aim of this series is to offer bite-size but comprehensive introductions, so here we've opted to merge both approaches.

Astronomy in Minutes, therefore, begins close to home with a look at the way in which our location on Earth shapes our understanding of the wider Universe, and the ways in which astronomy has developed to overcome some of the limitations of Earthbound observing. From there, we venture out among the objects of our own solar system – major planets, their large moons, and some more obscure but fascinating bodies.

Our first encounter with the wider Universe comes through an introductory guide to the 88 constellations that cover Earth's skies, and some of the most impressive objects within them. We then move on to look at the stars and associated objects in more detail, through the lens of the stellar life cycle. Breaking free of the Milky Way, we explore the realm of the galaxies and the large-scale Universe, before finally looking at cosmology – the science that attempts to answer some of the most fundamental questions about the origins and nature of the Universe, and which might have seemed strangely familiar to our prehistoric stargazing ancestors.

Discovering the Universe

Despite the lack of telescopes or other optical aids to improve their view of the sky, the earliest stargazers used a variety of ingenious instruments to measure the positions of stars and track the changing locations of planets. From these, they were able to develop remarkably sophisticated models of planetary motion that allowed them to predict celestial events. By the 16th century, such measurements became so accurate that they were capable of providing evidence for the Copernican theory of a Sun-centred Universe, and even for German astronomer Johannes Kepler's laws of planetary motion, still used to describe the Universe more than four centuries later.

The invention of the telescope at the beginning of the 17th century triggered another revolution in astronomy, as our understanding of the Universe was no longer limited by the frailties of human vision. Telescopes not only magnified our views of remote worlds and distant stars, but also allowed entirely new and unsuspected objects – ranging from new planets to star clusters, interstellar gas clouds and distant galaxies – to be observed for the first time. In the 19th century, astronomers seized on the new invention of photography, which allowed them first to simply make a more precise record of what they saw through their telescopes, and later (as films became more sensitive) to capture far more light, and far more detail, than an observer sitting directly at the eyepiece could ever hope to see. The ability to capture faint light using long exposures also allowed the techniques of spectroscopy (splitting the image of a light source into a spectrum of different colours) to be used for the first time – just as laboratory breakthroughs were confirming that spectral 'signatures' could reveal the secrets of an object's chemistry.

In recent years, the rise of satellite observatories, solid-state electronics and computing have transformed astronomy yet again. The door has been opened for a host of new observing techniques that extend to counting and manipulating the individual photons of light from distant galaxies at the edge of the observable cosmos. And in the future, the oldest science is sure to change again, offering yet more unexpected insights into the Universe around us.

Earth in space

Our view of the Universe is inevitably shaped by our location within it. While early astronomers believed that the Earth was the centre of everything and the stars, planets and other celestial objects all orbited around it, today we know the reality is very different – Earth is just one of eight major planets and countless other objects orbiting the Sun on elliptical paths. The other stars are so unimaginably distant that even the nearest barely shift their apparent position as we move through space (though the ability to measure this tiny shift, known as parallax, is in fact the keystone of our cosmic distance scale – see page 258). Thanks to Earth's rotation, the stars appear to move across the skies from east to west each day, while the Sun appears to drift slowly eastwards against the background stars as our planet moves along its orbit. The Sun completes a circuit of the sky once a year, following a track known as the ecliptic, and because the planets all orbit in more or less the same flat plane, they are also usually found on or close to the ecliptic.

The celestial sphere

Even though astronomers have long understood the reality of Earth's situation in space, it can still be useful, for practical purposes, to ignore the relative distances of celestial objects and treat them as if they are pinned to the surface of a spherical shell enclosing the Earth at an arbitrary distance.

This 'celestial sphere' forms the basis of the coordinate systems by which we measure the positions of objects in the sky. The north and south celestial poles are projections of Earth's own poles onto the sphere, while the celestial equator is a dividing line directly above Earth's own equator, splitting the sky into northern and southern hemispheres. Earth's rotation causes the celestial sphere to spin every 23 hours, 56 minutes about an axis that is fixed at the celestial poles. Because Earth's axis of rotation is tilted at an angle to the plane of its orbit, the line of the ecliptic (the Sun's yearly path against the background stars) is tilted at an angle of 23.5 degrees to the celestial equator, crossing it in two places.

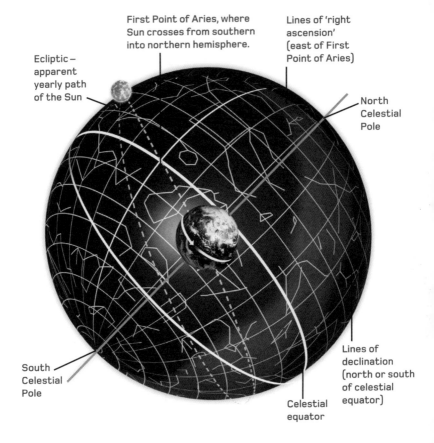

First Point of Aries, where Sun crosses from southern into northern hemisphere.

Lines of 'right ascension' (east of First Point of Aries)

Ecliptic – apparent yearly path of the Sun

North Celestial Pole

South Celestial Pole

Lines of declination (north or south of celestial equator)

Celestial equator

Celestial coordinates

Coordinate systems allow astronomers to define the location of objects in Earth's skies with reference to the imaginary celestial sphere (see page 12). The two most widely used systems are known as alt-azimuth and equatorial coordinates.

The alt-azimuth system is simply a measurement of an object's altitude (its angle above the horizon) and its azimuth (its direction, measured as an angle clockwise around the sky from due north). Unfortunately, the usefulness of these coordinates is limited – the sky's daily rotation means that the alt-azimuth coordinates of most objects are constantly changing, while the altitude and azimuth of the same object at the same time will vary from place to place. Equatorial coordinates are a more useful and portable system, albeit slightly more complex – they involve the measurement of properties called right ascension and declination (analogous to the longitude and latitude on Earth), which change only very slowly over time.

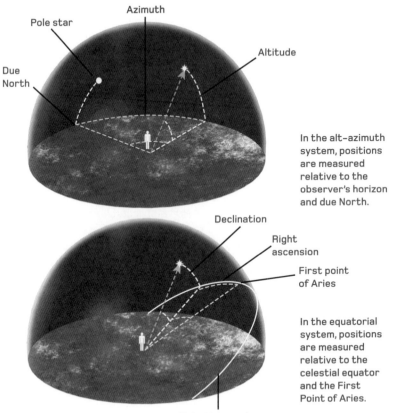

Pole star

Azimuth

Altitude

Due North

In the alt-azimuth system, positions are measured relative to the observer's horizon and due North.

Declination

Right ascension

First point of Aries

In the equatorial system, positions are measured relative to the celestial equator and the First Point of Aries.

Celestial equator

Electromagnetic radiation

The vast majority of our information about the Universe beyond Earth comes from the study of radiation from space – electromagnetic (e-m) waves that are emitted by all normal matter throughout the cosmos, whatever its temperature. These waves consist of interacting electrical and magnetic disturbances, aligned perpendicular to each other so that changes to the electric field reinforce the magnetic field and vice versa. This allows them to travel through space in self-propagating 'packets' called photons, moving at the speed of light (the ultimate speed limit of the Universe – about 300,000 kilometres or 186,000 miles per second).

The temperature of an object determines the energy of the radiation it emits, with higher energies corresponding to higher frequencies and shorter wavelengths. Only very hot objects such as stars have enough energy to emit visible light – cooler objects normally become visible by reflecting sunlight, although they emit invisible radiations of their own (see page 26).

A photon, or packet, of electromagnetic radiation consists of electrical and magnetic disturbances perpendicular to each other and vibrating at right-angles to the direction of their movement through space.

Telescopes

A telescope is essentially a device that uses a large surface area for collecting radiation (usually visible light) from distant objects and concentrating it on a detector of some sort (often simply a human eye). The result is a brighter image with higher resolution (more detail) than the detector alone could produce.

Visible-light telescopes come in two basic forms – refractors and reflectors. Refractors bend light to a focus as it passes through a transparent 'objective lens', while reflectors use a curved 'primary mirror' to bounce light rays onto converging paths. Both designs take advantage of the fact that light rays from distant objects are parallel to each other, and for direct observing purposes, both make use of a secondary lens or series of lenses to create a magnified image at the eyepiece. Invented around 1608, the earliest telescopes were refractors, as are many amateur instruments today, but the vast majority of modern professional telescopes are reflectors.

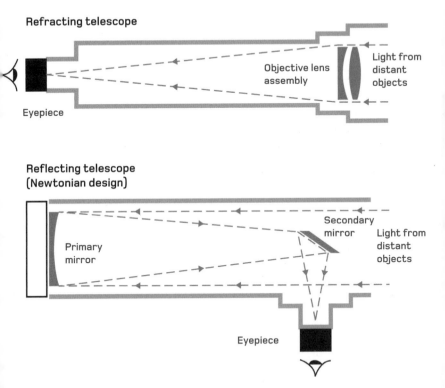

Refracting telescope

Eyepiece

Objective lens assembly

Light from distant objects

Reflecting telescope (Newtonian design)

Primary mirror

Secondary mirror

Light from distant objects

Eyepiece

Messengers from space

Most astronomers study the Universe via electromagnetic radiation (see page 16), but there are other ways of learning about the environment beyond Earth. Our planet is constantly bombarded by the solar wind (a stream of fast-moving particles blowing out from the Sun), and also by cosmic rays, which despite their misleading name are actually much faster-moving particles emitted by objects including exploding stars and giant black holes. Most of these particles do not make it to Earth's surface, but the 'air showers' of lower-energy particles they produce as they enter Earth's atmosphere can be identified using ground-based detectors.

Meteorites offer another important source of information about our solar system – these fragments of rocky debris can survive their fiery entry into Earth's atmosphere almost unscathed, and provide valuable samples not only of large asteroids such as Vesta (see page 60), but also of the raw material from which the solar system formed.

An early engraving depicts the spectacular Leonid Meteor storm that filled the skies of North America in November 1833.

Space probes

One important way of learning about astronomical objects is to visit them, and in the six decades since the dawn of the Space Age, we've done just that. While astronauts have so far travelled no further than Earth's Moon, robot spacecraft have ventured across the solar system to all the major planets and beyond. Space probes have trundled across the surface of Mars, braved the hostile atmosphere of Venus, roamed among the asteroid belt and flown alongside sputtering comets.

Probes not only send back close-up images showing more detail than could ever be obtained from Earth – they also measure planetary properties that are impossible to detect over interplanetary distances, such as the mineral compositions of rocky surfaces, the structure of magnetic fields and planetary interiors, and even the topography of the landscapes below. However, sending them into the hostile environment of the solar system is a hugely ambitious endeavour, and occasional failures are inevitable.

A self-portrait of NASA's Curiosity rover on the surface of Mars.

Modern astronomy

Astronomers today use a huge range of technologies to study the sky, including giant optical telescopes situated on mountaintops around the world, and a variety of instruments designed for looking at invisible radiations such as radio waves, the infrared and ultraviolet (see pages 26–33). Scientific researchers rarely look through instruments directly, relying instead on a range of detectors that collect and analyse electromagnetic radiations from the sky in different ways.

Detectors can range from simple digital CCD cameras, to photometers that measure an object's brightness by recording the number of photons striking the detector. Spectroscopes, meanwhile, which split an object's light up into a rainbow of colours in order to reveal the varying intensity of different wavelengths, are used to identify the chemical make-up of distant objects and discover other physical properties such as their motion through space (see page 264).

The invisible Universe

The visible light that provides us with our most immediate experience of the wider Universe is just a small part of the much wider spectrum of electromagnetic radiation ranging from long-wavelength, low-energy radio waves at one end, to high-energy, short-wavelength gamma rays at the other.

The full electromagnetic spectrum extends from radio waves and microwaves, through infrared to visible light, and beyond into ultraviolet radiation, X-rays and gamma rays. The

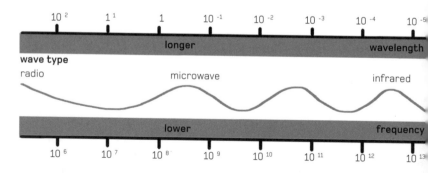

wavelengths involved range from millimetres to kilometres (in the case of radio waves) to quadrillionths of a metre (for gamma rays). Visible light runs from about 390 nanometres to 700 nanometres (billionths of a metre).

Astronomical objects such as stars emit radiation across a broad range of wavelengths, but the wavelengths tend to grow shorter (and the energies higher) with increasing surface temperature. Cool objects such as planets may only emit radiation in the infrared, while hot ones pump out most of their radiation as ultraviolet or even X-rays. Still other processes and objects, such as energized gaseous nebulae (see page 262) emit or absorb radiation at very specific wavelengths linked to their atomic structure.

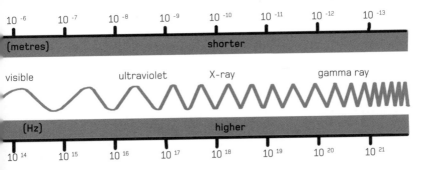

Radio astronomy

Many types of radio emissions from space arrive on Earth more or less unaffected by the atmosphere, but their very long wavelengths mean that, for a given size of telescope, radio images have much less resolution. In order to overcome this problem, radio astronomers build enormous dish-shaped receivers that focus waves onto an antenna where they are converted into electric signals as the dish scans across the sky. The huge size of individual telescopes helps make up for some of the resolution shortfall, while a technique called interferometry can combine signals from an 'array' of many telescopes to mimic the resolution of a much larger single dish.

Wavelengths between 1 metre and 1 millimetre are known as microwaves. The shortest of these are easier to image, but suffer heavily from absorption in Earth's atmosphere. In order to study microwave phenomena such as the afterglow of the Big Bang (see page 382), astronomers therefore launch microwave telescopes on satellites or high-altitude balloons.

Infrared astronomy

Infrared or heat radiation is emitted by many objects that are too cool to shine in visible light, as well as forming a significant part of the energy output for stars like our Sun, and even cooler red and orange stars (see pages 284 and 312). However, infrared signals from space are hard to detect because of both absorption and radiation by Earth's atmosphere. Mountaintop telescopes sited above the bulk of the air (and especially its water vapour) can observe the relatively high-energy near-infrared sky, but observations of the cooler far-infrared are only possible using orbital telescopes entirely above the atmosphere. Further complications are added by the tendency of such telescopes to detect their own infrared emissions – a problem that can only be overcome by cooling equipment to extremely low temperatures with liquefied gas. This has a tendency to slowly evaporate, limiting the lifetime of infrared satellites. The challenges are worth overcoming, however – the infrared is our only way to see an invisible cool Universe of planets, dwarf stars and interstellar dust.

A pair of images show the nearby Andromeda Galaxy (see page 160) in visible light and infrared radiation. Hot stars disappear in the infrared view, but the galaxy's skeleton of cool, dark dust becomes visible.

High-energy astronomy

Electromagnetic waves with shorter wavelengths and higher energies than visible light are divided into ultraviolet radiation, X-rays and gamma rays. The hottest stars emit more radiation in the ultraviolet than in visible light, while X-rays tend to come from superhot gas (for example, within galaxy clusters and around black holes), and gamma rays come from violent events that range from solar flares to supernovae. All three forms are largely blocked from reaching Earth's surface by the shielding effect of the atmosphere, and are best studied from satellites. Ultraviolet telescopes mostly follow the design of visible-light instruments, but the higher energies of X-rays and gamma rays present problems: they pass straight through most reflecting surfaces and so are hard to focus. Some X-ray telescopes get around this by 'ricocheting' rays at shallow angles off conical metal surfaces, but gamma ray telescopes, in particular, essentially rely on counting the rays that enter the 'window' of a heavily shielded detector as its scans different areas of the sky.

Gases in the expanding supernova remnant Cassiopeia A, heated to tens of millions of degrees, are best seen at X-ray wavelengths.

The solar system

By the most generous definitions, our solar system covers a huge volume of space extending up to 1 light year (9.5 trillion kilometres or 5.9 million million miles) from the Sun in every direction. This is the region of space in which the Sun's gravity holds sway, controlling the orbits of trillions of individual objects from giant planets to tiny fragments of cometary dust. While some astronomers prefer a stricter definition of the solar system as the much smaller region under the influence of the solar wind (also known as the heliosphere – see page 130), the gravitational definition ensures that the countless cold, dormant comets of the Oort Cloud, which began their lives much closer to the Sun and still follow long, slow orbits around it, are also included.

Distances within the solar system are typically measured in millions or even billions of kilometres – numbers that bear no relationship to our everyday experience. In order to simplify matters, astronomers therefore often refer to distances in our solar system and others in terms of 'astronomical units' (AU), where 1 AU is the average distance between Earth and the Sun – 149.6 million kilometres (93 million miles).

One key distinction among solar system objects is between major and dwarf planets. The eight major planets are worlds in orbit around the Sun, whose mass is large enough for gravity to pull them into a more-or-less spherical shape, and also to disrupt the orbits of anything else nearby, 'clearing out' a swathe of nearby space. Dwarf planets – a class introduced in 2006 following the discovery of Eris, but also encompassing Ceres, Pluto and others (see pages 124, 62 and 122) are spherical bodies orbiting the Sun that lack the gravitational punch to clear other objects from their neighborhood.

Four rocky planets orbit close to the Sun – Mercury, Venus, Earth and Mars – while four giant planets circle further out: the gas giants Jupiter and Saturn, and the ice giants Uranus and Neptune. Most have their own satellites (objects of varying size, complexity and origin held in orbit by a planet's gravity), and among and beyond them orbit a host of much smaller worlds, loosely referred to as minor planets. These include both rocky asteroids and icy comets (see pages 58 and 120).

Family of the Sun

1 Sun
Rotation period: 25–35 days
Diameter: 1,391,700 km
(864,400 miles)

2 Mercury
Orbital period: 88 days
Rotation period: 58.6 days
Diameter: 4,878 km (3,030 miles)

3 Venus
Orbital period: 224 days
Rotation period: 243 days
Diameter: 12,104 km (7,518 miles)

4 Earth
Orbital period: 365.25 days
Rotation period: 24 hours
Diameter: 12,756 km (7,923 miles)

5 Mars
Orbital period: 687 days
Rotation period: 24.6 hours
Diameter: 6,787 km (4,216 miles)

6 Ceres (dwarf planet)
Orbital period: 4.6 Earth years
Rotation period: 9.1 hours
Diameter: 975 km (606 miles)

7 Jupiter
Orbital period: 11.86 Earth years
Rotation period: 9.9 hours
Diameter: 142,800 km (88,700 miles)

8 Saturn
Orbital period: 29.5 Earth years
Rotation period: 10.6 hours
Diameter: 120,500 km (74,800 miles)

9 Uranus
Orbital period: 84.2 Earth years
Rotation period: 17.2 hours
Diameter: 51,118 km (31,750 miles)

10 Neptune
Orbital period: 164.8 Earth years
Rotation period: 16.1 hours
Diameter: 49,528 km (30,762 miles)

11 Pluto (dwarf planet)
Orbital period: 247.7 Earth years
Rotation period: 6.4 days
Diameter: 2,370 km (1,470 miles)

12 Eris (dwarf planet)
Orbital period: 560.2 Earth years
Rotation period: c.25.9 hours
Diameter: 2,400 km (1,490 miles)

The Sun

Our local star is a fiery ball of gas 1.39 million kilometres (863,000 miles) across. Powered by nuclear fusion, it has a three-layered internal structure (see pages 272 and 286). Its visible surface or 'photosphere' marks the zone where the Sun's hot gases grow sparse and cool enough to become transparent, at a temperature of around 5,500°C (9,900°F), but the Sun's sparse atmosphere or corona extends much further outwards. It reaches temperatures of more than a million degrees as it merges with the flow of the solar wind – a stream of particles blowing out across the solar system. The Sun's outer layers are constantly changing as its magnetic field varies in strength and complexity over an 11-year 'solar cycle'. Magnetic loops arcing high above the surface create dark, cool patches in the photosphere known as sunspots, and streams of dense gas in the lower corona called prominences. When they 'short circuit' closer to the surface, they release huge bursts of energy and torrents of particles known as solar flares and coronal mass ejections.

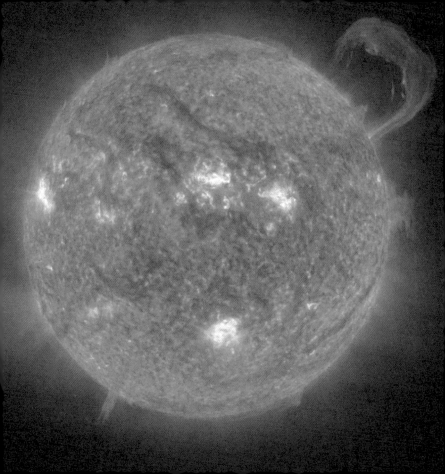

Mercury

The closest planet to the Sun is also the smallest of the major planets, with a diameter just 40 per cent bigger than Earth's Moon. Mercury hurtles around the Sun in just 88 days on a notably elliptical orbit, but slowed by tidal forces has a long rotation period of 57 days (two-thirds of a Mercury year). This means that the Sun moves very slowly across Mercury's skies, with an average interval of two Mercury years between successive sunrises. The long day, proximity to the Sun and lack of atmosphere combine to give Mercury extreme temperature variations, peaking at around 425°C (797°F) at midday, but plunging to -195°C (-319°F) on the night side.

Despite its small size, space probe investigations have shown that Mercury is surprisingly dense, with a weak magnetic field. Cracks and faults running across its heavily cratered surface suggest the planet has an unusually large core, which caused it to expand early in its history, cracking the crust apart before it shrank back to its present size.

Venus

Named after the classical goddess of beauty, Venus is the brightest object in Earth's skies after the Sun and Moon, but its name is deceptive – the second planet from the Sun is a hellish world cloaked in a choking, toxic atmosphere. Atmospheric pressure is 100 times that on Earth, with the air dominated by carbon dioxide. Sulfuric acid rains down from clouds, but evaporates before it can reach the ground. Despite a rotation period of 243 days that is actually longer than a Venusian year, the blanketing effect of the atmosphere ensures searing temperatures, surpassing 460°C (860°F), across both day and night sides. Considering its similarities to Earth, these stark differences are surprising. Most researchers believe they are linked to loss of water from the atmosphere early in Venusian history, and a subsequent runaway greenhouse effect. Radar mapping of the surface indicates that it has been extensively reshaped by volcanic activity that may still be going on today, but there is no sign of the fractured tectonic plates seen on Earth's surface.

This image of the volcanic Venusian landscape was reconstructed using radar data gathered through thick atmosphere by NASA's Magellan probe.

Earth

Planet Earth is the third from the Sun, and the largest of the solar system's rocky worlds. It circles the Sun at an average of 149.6 million kilometres (93 million miles) – a distance that defines the astronomical unit (AU), used as a convenient measuring scale for the solar system.

Earth's predominant features are abundant surface water, a crust split into tectonic plates and, of course, the presence of life. Today, scientists understand that all these features and many more are linked, affecting each other in unpredictable ways to produce the most complex planetary environment in the solar system. Earth's size means it has the hottest interior of any rocky planet – the transfer of heat from the molten core through the rocky mantle drives the slow motion of the tectonic plates at the surface. Meanwhile, Earth's gravity and location relative to the Sun allow water to exist as both solid, liquid and vapour. This unique combination of circumstances shapes the geology of Earth's surface and allows life to thrive.

Near-Earth space

The region of space around Earth is dominated by our planet's powerful magnetic field. Generated by swirling currents of molten iron in the outer core, the 'magnetosphere' resembles the field around a bar magnet, emerging from one magnetic pole, looping around the planet and re-entering at the other. The orientation of the magnetic poles is close to the axis of Earth's rotation, but the deep currents that generate it periodically reverse, causing the field to flip every few hundred thousand years.

Earth's magnetosphere plays an important role in protecting our planet from high-energy cosmic rays (see page 20) and charged particles from the solar wind. The majority are deflected by the field, but others are swept up and accelerated to even higher speeds, forming radiation belts around our planet. Particles filtering down from the magnetosphere above the poles collide with the gases of the upper atmosphere to form glowing aurorae – northern and southern lights.

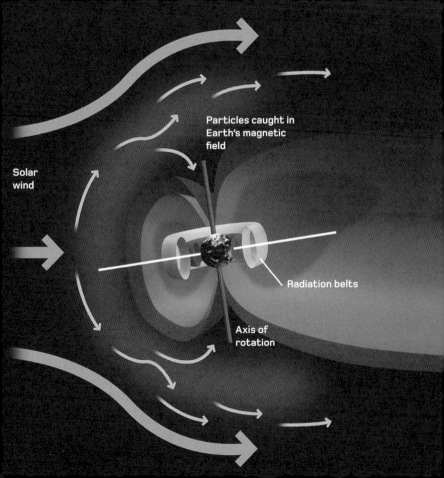

Solar wind

Particles caught in Earth's magnetic field

Radiation belts

Axis of rotation

The Moon

Earth's Moon is the most prominent object in the night sky, constantly varying in appearance as its illuminated area changes through each 27.3-day orbit. Synchronous rotation (in which the Moon also spins on its axis in 27.3 days) keeps one face permanently turned towards Earth while the other faces permanently away. The near side is divided into two basic types of terrain – bright, heavily cratered highlands and darker, smoother 'seas' or 'maria' – while the far side is almost entirely composed of highland terrain, with only a handful of maria. The seas are, in fact, ancient lava plains, created more than 3 billion years ago when volcanic activity flooded the low-lying basins left behind by major impacts on the near side. The lack of seas on the far side, in contrast, may be due to the fact that the crust here is significantly thicker (perhaps in turn due to tidal forces, see page 50), and so molten magma found it more difficult to reach the surface. Since the eruptions that formed the maria, the Moon has been geologically dead, though craters have continued to accumulate on its surface.

The Earth-Moon system

Earth's lone satellite is so big compared to its parent planet that Earth and Moon exert considerable influence on each other and some astronomers consider them as effectively a 'double planet'. The most obvious influence is in the form of tides – effects caused by the substantial variation of each body's gravitational attraction as felt throughout the other. On Earth, tides are made more obvious by the presence of deep oceans, allowing the Moon to pull a metres-high 'tidal bulge' towards it, and creating a slightly smaller bulge on the opposite side of the planet. As Earth rotates beneath the Moon each day, most coastal areas experience a cycle of two high and two low tides. Tidal forces from Earth influencing the Moon are also responsible for slowing lunar rotation to its current synchronous rate (matching its orbital period), but are also causing it to slowly spiral away from Earth by a few centimetres each year. Today, the Moon experiences only slight tidal forces as its distance from Earth varies throughout each month, creating periodic 'moonquakes'.

Near-Earth objects

The space around Earth is filled with objects on a variety of scales, following their own orbits around the Sun. The largest are near-Earth asteroids (NEAs) – rocky bodies mostly kicked out of the Kirkwood gaps in the main asteroid belts (see page 60) into elliptical orbits that come close to the inner planets. NEAs are broadly divided into four classes, each named after the first of their type to be identified: Amor asteroids have orbits entirely beyond that of Earth; Atiras or Apophele asteroids have orbits entirely inside Earth's; while Apollos and Atens are asteroids in Earth-crossing orbits with long axes longer or shorter than 1 AU respectively (though tilted orbits mean that most do not represent a threat to Earth).

Other significant near-Earth objects include comets and their associated debris streams – small fragments of ice and dust left behind by the passage of comets. When our planet crosses these debris streams, the fragments burn up in Earth's atmosphere as periodic shooting stars or meteor showers.

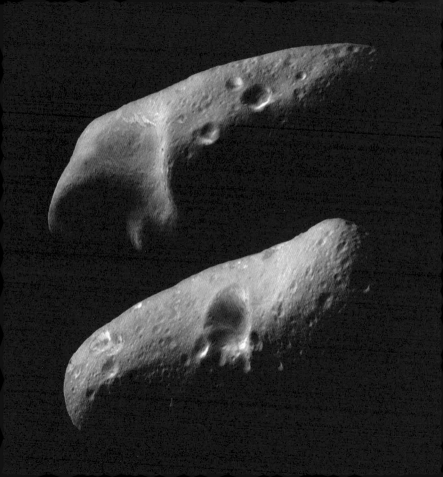

Mars

The famous Red Planet is a cold, dry world, considerably smaller than Earth, but is still the most hospitable body in the rest of the solar system. The planet gets its colour from dust rich in iron oxide (rust) that covers much of its surface. Its crust is divided into a heavily cratered southern highlands, broad northern plains and a huge bulge in the crust known as the Tharsis Rise, which plays host to many of the largest volcanoes in the solar system. Elsewhere, a huge valley system known as the Valles Marineris dwarfs Earth's Grand Canyon.

Mars follows a notably elliptical orbit around the Sun (see page 36), and has an orbital tilt similar to Earth's, giving it a complex pattern of seasons. Its atmosphere exerts a pressure just 0.6 per cent of Earth's, and is dominated by carbon dioxide, which forms extensive frost caps over the poles during winter. However, both poles are underlain by huge deposits of water ice in a deep-frozen permafrost, and there is plentiful evidence that water once flowed freely across the surface of Mars.

The Martian mountain Olympus Mons is the largest volcano in the solar system, roughly 600 kilometres (370 miles) in diameter.

Phobos and Deimos

Mars has two small moons, named after the sons of the Greek war-god Ares. Phobos has dimensions of 27 x 22 x 18 kilometres (17 x 14 x 11 miles) and orbits in just 7 hours 39 minutes – much less than a Martian day. Its surface is dominated by a large crater, called Stickney, and covered in curious parallel scars. Deimos is smaller and more distant, with an orbital period of 30 hours 19 minutes. In contrast to the heavily cratered Phobos, its surface seems to be smoothed by large amounts of dust.

Astronomers used to assume that the moons were asteroids captured into Martian orbit, but recent calculations suggest that such an event (let alone two) would require a very unusual set of circumstances. An alternative idea is that they formed in a similar way to Earth's Moon, coalescing out of debris ejected by a major impact on the planet's surface. One thing that is certain is that tidal forces will eventually cause Phobos to shatter into fragments and rain down on the Martian surface.

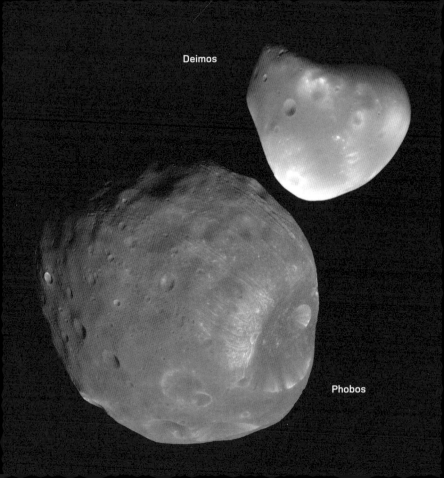

Comets

Icy chunks of debris from the solar system's cold outer wastes, comets can become occasional spectacular visitors to Earth's skies. Normally they orbit in the remote Kuiper Belt and Oort Cloud, but when their paths through space are disrupted by a collision or gravitational influence, they may fall towards the Sun on narrow elliptical orbits. While a comet's complete orbit may take centuries or even millennia, its journey through the inner solar system and around the Sun typically lasts just a few months. During this 'perihelion passage', heat from the Sun warms the comet's interior, causing ice to evaporate and erupt through the surface. The result is a huge but tenuous extended atmosphere called a coma, surrounding a solid 'nucleus' that is typically just a few kilometres across. Ionized (electrically charged) gas and dust caught up on the solar wind stream away from the nucleus to form beautiful tails, glowing through both emission and reflection of radiation. Each passage around the Sun burns off a little more of the comet's ice, giving it a limited lifetime before it is exhausted.

Orbit of Earth

Developing tail

Nucleus

Ion tail points
directly away
from Sun

Inner part of
long elliptical
comet orbit.

Comet
approaches
Sun

Sun

Dust tail curves
along orbit

Comet becomes
less active as it
retreats from Sun

Main-belt asteroids

Between the orbits of Mars and Jupiter lies the solar system's main asteroid belt – a region of rocky debris that persists where the gravitational influence of Jupiter prevented material accumulating to form a fifth rocky planet. Many millions of individual objects orbit here, but their combined mass is only about 4 per cent of Earth's Moon – the vast majority of rocky debris that once filled this region was ejected by interactions with Jupiter in the early days of the solar system. The surviving asteroids include some 200 with diameters of more than 100 kilometres (61 miles), but just four larger bodies – the dwarf planet Ceres and the asteroids 4 Vesta, 2 Pallas and 10 Hygiea – account for more than half of the belt's mass.

Main-belt asteroids follow a variety of orbits around the Sun, with different eccentricities and inclinations, but a few distinct orbits known as the 'Kirkwood gaps' are empty. Here, the repeated influence of Jupiter's gravity can still kick an asteroid out of the belt, perhaps turning it into a near-Earth object.

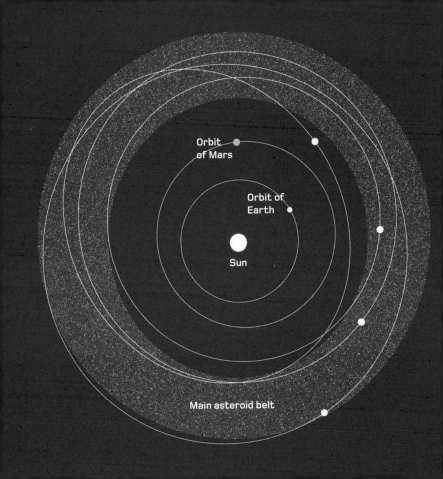

Ceres

By far the largest resident of the asteroid belt with a diameter of some 950 kilometres (590 miles), Ceres has officially been classed as a dwarf planet since 2006. This is defined as a body in an independent orbit around the Sun, with enough gravity to pull itself into a spherical shape but not enough to clear other objects from the space around its orbit.

While it was long suspected to be nothing more than a particularly large chunk of rock, there is increasing evidence that Ceres is rather deserving of its planetary status. Images from the Hubble Space Telescope have shown a distinct reddish surface and bright patches thought to be surface water ice, while spectroscopic studies have pointed to the presence of a thin atmosphere, potentially containing water vapour. Ceres certainly contains unexpected amounts of ice, and some researchers have even speculated that Ceres might have a hidden ocean of liquid water beneath its surface, rather similar to that on Jupiter's moon Europa.

Vesta

As its official designation suggests, Vesta was the fourth asteroid to be discovered (in 1807), though it is actually the brightest member of the asteroid belt and often just visible to the naked eye. It is about 525 kilometres (326 miles) across, and would be large enough to pull itself into a spherical shape were it not for the presence of a huge impact crater, known as Rheasilvia, which takes a huge gouge out of Vesta's south pole.

Vesta was targeted for a year-long investigation by NASA's Dawn space probe in 2011, whose observations suggested that Vesta has a differentiated interior with an iron core, and was geologically active in its distant past. Many Vesta-like asteroids formed early in the solar system's history, but most have either been expelled from the main belt or broken up by collisions. Experts believe that fragments from different parts of their interiors may have given rise to distinct families of iron, stony and stony-iron meteorites collected on Earth, some of which have been conclusively linked to the crust of Vesta.

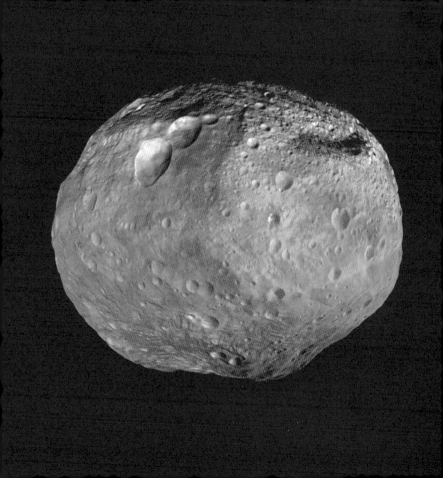

Jupiter

The biggest planet in the solar system, Jupiter is large enough to swallow up all the others with room to spare. It is a gas-giant world, dominated by a huge envelope of hydrogen gas (compressed into liquid some 1,000 kilometres or 600 miles below the surface) around a solid core that may be no larger than Earth. At great depths, the liquid hydrogen breaks up into individual atoms to create an electrically conductive 'liquid metal' form. Swirling currents in this layer generate a powerful and extensive magnetic field that makes its presence felt as far away as Saturn, while the slow contraction of the planet's inner layers, driven by its enormous gravity, releases just as much heat as Jupiter receives from the Sun. Seen from a distance, however, the planet's dominant features are its colourful banded weather systems, broadly divided into pale 'zones' and dark 'belts' running parallel to the equator. The Great Red Spot, an enormous storm in the southern hemisphere, is about twice the diameter of Earth and has been observed continuously for almost two centuries.

The Jovian system

Jupiter's powerful gravitational field dominates a huge area of space, and puts it at the heart of a system of at least 67 moons. Eight of these are thought to be 'natural' satellites, formed from material left in orbit around Jupiter after the planet's creation, while the remainder are 'irregular' moons – minor planets that have been captured into orbit around Jupiter in the billions of years since its origin. Most of the irregular moons belong to one of four major groups linked to the break-up of larger progenitor worlds. Each contains a single moderate-sized satellite (Himalia, Carme, Ananke and Pasiphae) and a number of smaller bodies in similar orbits.

The natural satellites fall into two groups: four small inner moons – Metis, Adrastea, Amalthea and Thebe – and four planet-sized 'Galilean' moons – Io, Europa, Ganymede and Callisto. Micrometeorite impacts on the surfaces of the inner moons eject material into orbit around Jupiter, creating a thin, dusty ring system that is invisible from Earth.

Jupiter's volcanic moon
Io, seen against the cloud
belts of the giant planet.

Io

The innermost of Jupiter's Galilean moons, Io orbits Jupiter in just 42.5 hours, so close to its parent planet that it is bombarded by dangerous radiation in Jupiter's magnetic field. It is also pummelled in a tidal tug-of-war between the giant planet on the one side and its outer neighbours Europa and Ganymede on the other. These tidal forces generate huge amounts of heat in Io's sulfur-rich rocks, creating pockets of molten magma that erupt at the surface to make Io the most volcanically active world in the solar system. Huge plumes of liquid sulfur and sulfur dioxide rise up to 500 kilometres (300 miles) above the surface before falling down to blanket the landscape in colourful red, orange, white and yellow sulfur compounds. Elsewhere, molten sulfur oozes directly onto the surface through volcanic vents, while the sheer weight of material being deposited on the surface causes the crust to buckle in places, pushing up towering mountain ranges. All this activity means that Io's pockmarked face is constantly evolving and few features persist for more than a few decades.

Europa

The second of the Galilean moons is also the smallest, but perhaps the most intriguing of all. From a distance, Europa appears as a remarkably smooth, brilliant-white ball of ice, but enhanced images reveal that it is criss-crossed by pale markings and has very few impact craters. Such features suggest that Europa is constantly resurfacing itself.

Europa's orbit between Io and Ganymede renders it vulnerable to the same tidal heating that shapes Io, but the presence of large amounts of water ice means that the heat has a rather different effect: the moon has separated into a rocky core and an icy crust, with a deep liquid-water ocean between them. Tidal heating of the core helps keep the ocean warm, while flexing and cracking of the surface allows fresh water or slushy ice to well up from below, sometimes escaping in jets of water arcing high above the surface. Seafloor volcanoes may even pump chemicals into the ocean, making Europa perhaps the most suitable place for life elsewhere in the solar system.

Recent impact craters leave bright 'splashes' on Europa's icy surface.

Ganymede

The largest moon in the solar system, Ganymede is the only satellite with a substantial magnetic field of its own, suggesting that its interior has fully separated to produce a core that still contains a significant amount of molten iron. Magnetic evidence also points to a saltwater ocean layer some 200 kilometres (125 miles) below the surface. The crust, meanwhile, has a mottled appearance with two distinct types of terrain: dark, old heavily cratered regions separated by lighter, less cratered (and therefore more recently formed) areas that are often laid down in parallel grooves and ridges.

Unlike its inner neighbours Io and Europa, this giant moon does not experience major tidal forces today, but there is strong evidence that it once did. Tidal forces, combined with heat left over from Ganymede's formation, for a time drove an icy equivalent of Earth's plate tectonics, with older and darker areas of the crust pulled apart and a fresh rock-ice mix welling up from within to fill the gaps.

Callisto

The outermost and second largest of Jupiter's Galilean moons, Callisto is very different from its inner neighbours. Its dark surface is scarred by countless craters, the most recent of which have sprayed bright rays of icy ejecta over the surrounding landscape. Too far from Jupiter to experience significant tides, Callisto's interior never separated into distinct layers, and instead remains a largely undifferentiated mix of rock and ice (though there is some evidence for a possible saltwater ocean layer deep beneath the surface). As a result, the moon's terrain has never been resurfaced, and it has been shaped only by the constant bombardment of meteorites pulled towards Jupiter by its powerful gravity and the curious erosive power of the Sun. Particles from the solar wind cause chemical reactions that slowly darken the surface, while heat from its weak radiation causes ice to sublimate out of the rock-ice mix, weathering the crater walls into chains of jagged peaks. Small wonder that some have called Callisto the most densely cratered world in the solar system.

Trojan asteroids

Since the early 20th century, astronomers have been aware that Jupiter shares its orbit with several thousand asteroids (almost 6,000 at the latest count). These small bodies avoid the dangerous influence of their neighbour's powerful gravity by orbiting in clouds 60 degrees behind and ahead of Jupiter itself. These two so-called Lagrangian points are gravitational 'sweet spots' where the Sun's influence cancels out that of Jupiter and a stable orbit is possible.

The asteroids in these two clouds have traditionally been named after soldiers from the rival sides of the mythological Trojan War, and are collectively known as the Trojans (though about half of them represent figures from the Greek side). By extension, any object in this kind of orbit has also become known as a Trojan, so Mars, Uranus and Neptune all have Trojans of their own. Earth's first Trojan, the asteroid 2010 TK$_7$, was identified in 2011. Saturn's moons Tethys and Dione, meanwhile, share their own orbits with Trojan satellites.

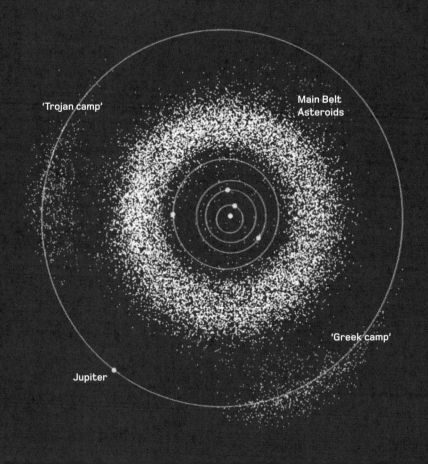

Saturn

Saturn is the most distant of the naked-eye planets known to the ancients, and the second largest planet in the solar system. But despite having almost 60 per cent of Jupiter's volume, it has less than a third of its mass, and so is the least dense planet in the solar system (Saturn is less dense than water). In many other respects, Saturn is similar to Jupiter – a gas giant with a small core, dominated by hydrogen in gaseous, liquid and metallic forms.

Saturn's outward appearance, however, is rather different, because of the lower temperature of its upper atmosphere (arising from Saturn's low density and greater distance from the Sun). This allows a layer of creamy-white ammonia clouds to envelope the planet at high altitudes, muting most of the colour in the deeper cloud belts below. Saturn has no long-lived features equivalent to Jupiter's Great Red Spot, but it does have large vortices at the poles, and huge white storms that erupt regularly at mid-latitudes during northern summer.

Saturn's rings

All the giant planets are now known to have ring systems, but Saturn's are by far the most spectacular. They were first observed in 1610, but it was not until 1859 that Scottish physicist James Clerk Maxwell explained their true nature.

From a distance, the rings appear to be divided into several broad loops – the bright A and B rings, the fainter C and D rings nearer the planet, and so on, with dark gaps such as the Cassini and Encke divisions separating them. Up close, however, the rings are revealed to be composed of countless slender ringlets, each of which itself comprises millions of individual

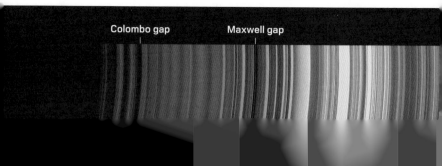

ring particles following perfectly circular orbits in a narrow plane above Saturn's equator (in order to avoid collisions with one another). The variation in appearance of the major rings is largely due to differences in the size and appearance of the particles within them – they range from house-sized boulders in the A and B rings, through centimetre-scale particles in the C ring, down to tiny grains in the D and E rings.

There are various theories to explain the origin of the rings – one of the most popular is that they were formed during the relatively recent breakup of a comet or small icy moon, and will steadily degrade over time to a sparser appearance like those of the rings around Jupiter, Uranus and Neptune. However, there is also evidence that the ring material is actually quite ancient, and somehow regenerates and retains its fresh appearance through repeated collisions.

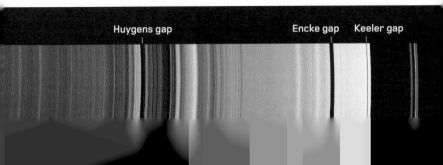

Huygens gap

Encke gap   Keeler gap

The Saturnian system

Saturn's system of moons rivals that of Jupiter, with some 62 known independent satellites, and many more small 'moonlets' orbiting within and around its rings. Of the known moons, some 44 are irregulars – captured asteroids or comets orbiting far from Saturn, in groups that may be associated with the break-up of a handful of larger objects. The remaining 18 closer satellites are natural moons, formed from debris left behind after Saturn's own formation. These moons are generally icier than the Galilean moons of Jupiter, and are much less evenly sized – the giant satellite Titan dominates the entire system, and even its largest neighbours are puny in comparison. Several small worlds orbiting close to Saturn act as 'shepherd moons' – objects whose gravitational influence herds ring particles into specific orbits, and which in some cases add material to the rings through tiny impacts that chip ice off their surfaces. Elsewhere, gravitational effects of the large outer moons open up gaps in the ring system where particles cannot maintain stable orbits.

Mimas

With a diameter of just 400 kilometres (250 miles), Mimas is the innermost of Saturn's larger natural satellites, and the smallest known body in the solar system with enough gravity to pull itself into a spherical shape (thanks in part to a high ice content). The result is a heavily cratered world that bears an irresistible resemblance to the *Death Star* space station from the *Star Wars* movies, thanks to the presence of the 130-kilometre (80-mile) crater Herschel. This crater is about as large as it's possible for an impact to get without threatening the integrity of the target object, and one theory for the origin of Saturn's rings is that they originated from the cataclysmic destruction of an icy moon that was not so lucky.

Despite the generally heavy cratering on the surface of Mimas, researchers have detected variations in crater density between different areas, suggesting that at least some areas experienced resurfacing early in the moon's history – most likely due to 'cryovolcanic' eruptions of a slushy rock-ice mix.

Enceladus

Saturn's second major moon is just slightly larger than its inner neighbour Mimas, but offers a stark contrast in other respects. Enceladus' surface is brilliant white, and huge geyser-like plumes of ice crystals that spray from deep fissures around the south pole, settling back onto the landscape as snows, soften the appearance of underlying craters. Not all the ice remains trapped by the moon's gravity – considerable amounts escape into orbit around Saturn, where it forms the faint, doughnut-shaped E ring.

Enceladus' orbit puts it in the middle of a gravitational tug-of-war between Saturn and the larger satellite Dione, and this warms the interior of this surprisingly rocky moon sufficiently to melt pockets of ice. The process is helped by the presence of ammonia alongside the water ice acting as an 'antifreeze' and lowering the melting point of the mix. Nevertheless, Enceladus is one of the most promising places in the solar system to look for extraterrestrial life.

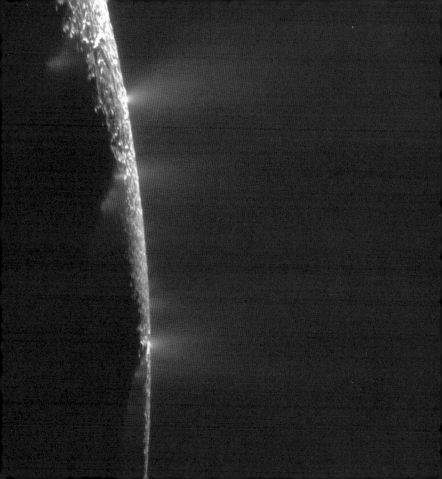

Tethys

Just as Mimas and Enceladus are almost identical in size, the next two major Saturnian moons, Tethys and Dione, are also near-twins, though considerably larger than their inner neighbours. Each is also accompanied by a pair of small Trojan moons sharing the same orbit (see page 78) – in Tethys' case these are Telesto and Calypso.

Tethys is bright but heavily cratered, with several large impact basins, the biggest of which is known as Odysseus. Despite their size, these basins are remarkably shallow, suggesting that the moon's surface has 'slumped' gradually over time. Tethys shows considerable signs of cryovolcanic activity (eruptions of slushy ice kept runny by the presence of ammonia) in its distant past. Measurements of its density, meanwhile, suggest that it is made almost entirely of water ice, and a huge canyon known as Ithaca Chasma, which runs across much of the moon's surface, probably formed as the planet's interior gradually froze solid and expanded.

Dione

Just slightly larger than its inner neighbour Tethys, Dione is accompanied in its orbit by a pair of Trojan satellites known as Helene and Polydeuces. It is considerably denser than Tethys, suggesting a higher proportion of rock in its interior, and shows variations in the number of craters across its surface, indicating that its crust was resurfaced by large-scale cryovolcanism in its past.

Dione shows several significant differences between its leading and trailing hemispheres, with the heaviest cratering on its forward-facing side, suggesting that it once flew through a veritable blizzard of debris (perhaps created during the catastrophic destruction of another moon). The darker trailing hemisphere, meanwhile, is criss-crossed by what appear to be networks of fine bright lines, nicknamed 'wispy terrain'. It was only in 2004 that NASA's Cassini probe revealed these bright lines to be the smooth, exposed faces of cliff-like geological faults scattered across the hemisphere.

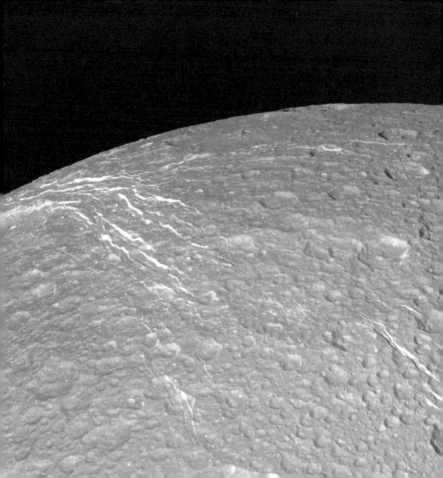

Rhea

At first glance, Rhea appears to be a larger version of Dione, with similar features including heavy cratering on its leading hemisphere, and a darker trailing hemisphere criss-crossed by meandering ice cliffs that give a cobweb-like appearance. Rhea also represents another step up in size across Saturn's system of moons – its diameter of 1,528 kilometres (949 miles) is second only to the giant satellite Titan, and just slightly larger than the more distant Iapetus.

Closer inspection, however, reveals some features unique to Rhea. Its craters are more sharply defined than those on Tethys and Dione, suggesting that its surface is less prone to 'slumping' – one theory is that Rhea's higher mass and gravity caused its ice to compress into a more rigid, less fluid form. Despite Rhea's heavily pockmarked overall appearance, an absence of large craters in certain areas suggests that they were wiped away by cryovolcanism very early in the moon's history – but Rhea then appears to have rapidly frozen solid.

A Cassini space probe image shows Rhea
and the smaller, more distant Epimetheus,
against ring shadows on Saturn.

Titan

Saturn's largest moon is one of the most complex and mysterious worlds in the solar system – uniquely cloaked in a thick, dense, nitrogen-rich atmosphere. Orange methane clouds concealed its secrets from early space probes, but infrared cameras and radar imaging by NASA's Cassini mission have now penetrated the haze to reveal a mix of raised highlands and lowland plains reminiscent of our own planet's continents and seabeds. What's more, the landscape bears the unmistakable signs of erosion by flowing liquid – and yet Titan's average surface temperature is a chilly -180°C (-292°F).

It seems that Titan's unique conditions allow hydrocarbon chemicals, such as methane, to exist in solid, liquid and gaseous forms, creating a 'methane cycle' similar to Earth's own water cycle. The transition between gaseous methane, liquid methane rain and frozen methane ice shapes the surface just as the flow of water does on Earth, and standing lakes of liquid methane have even been found at both of Titan's poles.

A sequence of mosaic images captures the view from the European Space Agency Huygens lander during its descent to Titan's surface in 2005.

Hyperion

Orbiting some way beyond Titan, Hyperion is a puzzling world that may point to dramatic events in Saturn's distant past. At 360 kilometres (224 miles) long, it should be large enough to pull itself into a sphere, but is unevenly shaped. Its rotation varies chaotically, and its surface has a strange, sponge-like structure, with deep pits separated by razor-sharp ridges.

The most plausible explanation for these curious features is that Hyperion is the surviving remnant of a larger moon that once orbited here before being shattered in a cataclysmic impact. The destruction of a large moon would have sent debris spiralling towards Saturn, where it was swept up by Titan, Rhea and Dione, explaining their distribution of craters. It may even have enriched Titan's atmosphere with volatile chemicals released by impacts. The surviving fragment of Hyperion ended up with its current unpredictable spin, while erosion of dark areas by solar radiation is probably responsible for shaping its pitted landscape.

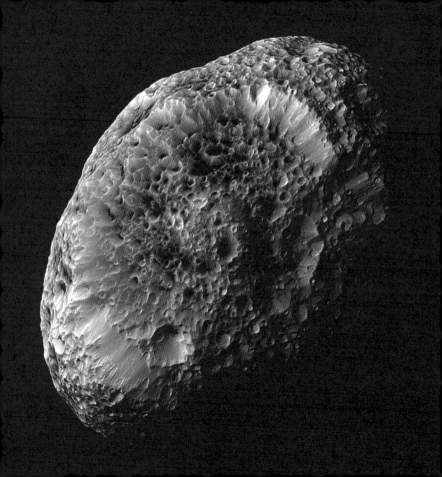

Iapetus

Iapetus is one of the strangest moons in the solar system – a world roughly the same size as Rhea, with a stark difference in brightness between its two halves. The trailing hemisphere, which faces backwards along the moon's orbit, has a similar bright, icy surface to inner moons such as Rhea and Tethys, but the leading hemisphere is extremely dark. Astronomers think that the difference arose as a result of a 'feedback loop' triggered by dust in the so-called 'Phoebe Ring' (see page 102) accumulating on the moon's leading face. The dust makes the surface slightly darker, and therefore more easily warmed by radiation from the distant Sun. This causes ice to sublimate out of the rock-ice mix that makes up the surface of Iapetus, leaving behind a dark, dusty 'lag' that increases the absorption of heat and drives off still more of the ice. While this mechanism is fairly well understood, other aspects of Iapetus – such as its inclined orbit relative to Saturn's other regular moons and the curious line of mountains that run directly around the moon's equator – are still not fully explained.

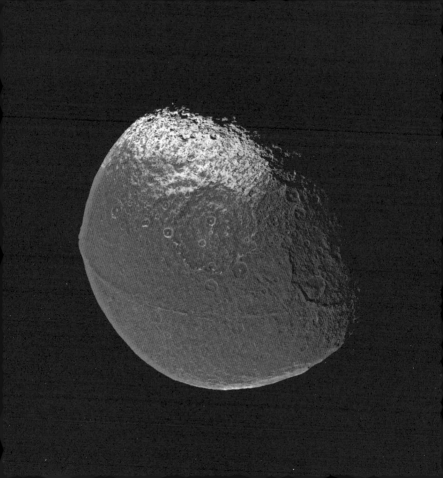

Phoebe

The innermost of Saturn's irregular moons is also the largest and the best understood, thanks to a 2004 flyby as NASA's Cassini space probe approached the ringed planet. Phoebe is a dark, unevenly shaped and heavily cratered little world, measuring 219 kilometres (136 miles) on its longest axis, and bears a strong resemblance to comets visited elsewhere in the solar system, although it is much larger. Cassini's images confirmed what had long been suspected – Phoebe is almost certainly a captured centaur, one of a group of small icy worlds that roam among the gas giants (see page 104).

Phoebe orbits Saturn on a notably elliptical and inclined path that is also 'retrograde' – in other words, it goes around Saturn in the opposite direction to the planet's rotation and that of its natural moons. In 2009, astronomers identified a broad ring of dust around Phoebe's orbit. This is probably created as micrometeorites bombard its surface, and is linked to the unusual surface of its inner neighbour Iapetus.

Centaurs

The space between the orbits of the giant planets is not empty. Small icy worlds wander between them on more or less elliptical orbits that are unstable and prone to disruption on timescales of millions of years or less. These objects, known as centaurs, were first recognized after the discovery of minor planet 2060 Chiron in 1977, and the entire population is thought to number perhaps 44,000 objects larger than 1 kilometre (0.6 miles) across. Earth-based observations have revealed surface colours ranging from red to blue, and spectral analysis has detected the presence of various carbon-based organic compounds. Several centaurs, including Chiron, have been seen to develop comet-like comas when at their closest to the Sun and the largest, 260-kilometre (161-mile) Chariklo, even has its own system of rings. The bulk of evidence suggests that centaurs are objects that originate in various parts of the Kuiper Belt (see page 120), and spend a short time in this region of the solar system before becoming fully fledged comets.

Uranus

The first planet to be discovered in the age of the telescope (in 1781), Uranus is actually bright enough at times to be seen with the naked eye. It is a pale blue world, considerably smaller than Jupiter or Saturn and with a rather different composition, shared with Neptune (see page 114). Uranus' most unusual characteristic is the extreme tilt of its axis of rotation – its north pole points down at an angle of 98 degrees relative to the plane of its orbit. As a result, the planet experiences extreme seasonal changes during its 84-year orbit around the Sun – high latitudes experience a long period of constant sunlight or darkness, and only regions close to the equator go through a day/night cycle linked to the planet's 17.2-hour rotation period. The extreme temperature variations created around midsummer/midwinter create strong winds between the poles, which tend to suppress other weather systems and leave the planet looking dull and placid (as during Voyager 2's flyby of 1986). In spring and autumn, however, the more normal weather allows cloud bands to develop and storms to flare up.

A recent Earth-based view reveals weather systems invisible at the time of the Voyager 2 mission's encounter with Uranus.

The Uranian system

Uranus is surrounded by a system of narrow rings that give the planet the appearance of a bullseye when viewed 'head on'. The 13 known rings are considerably darker than those around Saturn, and are thought to contain significant amounts of carbon-rich chemicals, such as methane, as well as water ice. Most of the rings are composed of fairly large particles with relatively little dust between them.

There are 13 small moons orbiting among the rings, acting as shepherds to keep them in their narrow tracks. Beyond these lie five 'major moons' – Miranda, Ariel, Umbriel, Titania and Oberon. All but Umbriel show signs of cryovolcanic resurfacing in their past, due either to the release of heat built up in their formation (in the case of the larger Titania and Oberon), or tidal forces. Nine small outer moons, meanwhile, are captured comets or asteroids trapped in irregular orbits. Uniquely, the moons are named after characters from the works of Shakespeare and Alexander Pope, rather than mythology.

A Voyager space probe image captures star trails passing behind the thin Uranian ring system.

Miranda

Tiny Miranda is the innermost of five Uranian moons known before the Space Age. With an average diameter of just 472 kilometres (293 miles), it was expected to have changed little from its formation, but images returned by the Voyager 2 space probe in 1986 revealed a surprisingly complex world, with a mix of heavily cratered and smoother terrain, separated by deep canyons and towering cliffs. Perhaps the most impressive features, however, are racetrack- or chevron-shaped regions covered in parallel grooves, known as coronae.

At first, astronomers wondered whether the small moon had shattered completely in its distant past, and then reassembled itself. Today, it's thought more likely that Miranda was shaped by extreme tides during a period of its history when the moon followed a much more eccentric orbit than it does today. Tidal heating melted the moon from the inside out, and large areas of its crust subsided, with new material welling up to replace them.

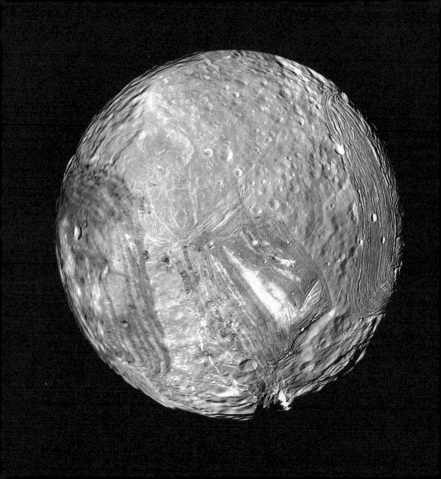

Ariel

The second major moon of Uranus displays a complex terrain formed during an active past. It has the brightest surface of any Uranian moon, generally grey in appearance but with a slight reddening of its leading hemisphere. This effect was probably caused by chemical changes due to bombardment from particles in the planet's magnetosphere.

Measurements of Ariel's density suggest that it is composed of a roughly even rock-ice mix, while its surface varies from cratered uplands to low-lying, less cratered plains, and areas of banded ridges and troughs. Even on the uplands, the overall level of cratering is less than would be expected if the landscape had been soaking up impacts for the entire history of the solar system, which suggests that the whole moon was resurfaced by icy cryovolcanic activity at some point in its youth, most likely due to tidal heating. Today, Ariel experiences only weak tides, but it once followed a different orbit that brought it under the influence of the outer moon Titania.

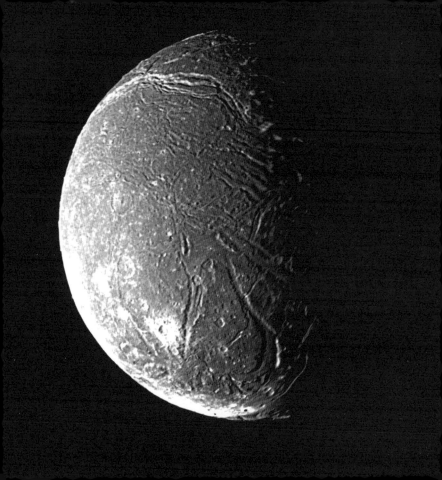

Neptune

The outermost of the solar system's major planets, Neptune is slightly smaller but has a larger mass than Uranus, with a deeper blue colour. Like its inner neighbour, it is an 'ice giant' – its interior is dominated by a mix of relatively complex chemicals such as water, methane and ammonia (all classed as ices in chemical terms). These are less volatile than the lightweight gases hydrogen and helium that dominate Jupiter and Saturn, and give the more distant planets a 'slushy' interior. Nevertheless, hydrogen and helium dominate their atmospheres, with small proportions of methane that absorbs red light from the Sun and gives the planets their bluish tints.

Despite its location in the cold depths of the outer solar system, Neptune has surprisingly active weather systems, with some of the fastest winds seen on any planet, and large storms that appear as dark spots. All this activity seems to be driven by a powerful internal heat engine that generates 2.6 times as much energy as Neptune receives from the Sun.

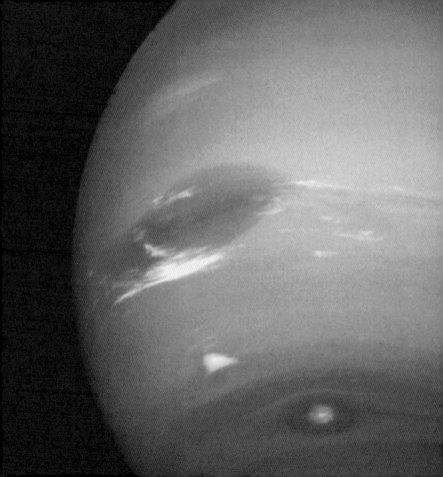

The Neptunian system

Neptune has a ring system similar to that of Uranus, but less substantial – the distribution of ring particles within them is notably uneven, and the outermost Adams ring displays five distinct clumps or arcs of material. Observational evidence suggests the rings are unstable and likely to decay in a matter of centuries.

Neptune's 14 known moons include seven small inner satellites orbiting within and just beyond the rings. They are believed to have originated from debris left behind after the destruction of an earlier generation of original satellites – a catastrophic consequence of the arrival of the largest Neptunian satellite, Triton, into orbit. Triton, believed to be a captured 'ice-dwarf' world (see pages 120-125), is today by far the largest moon of Neptune, following a perfectly circular orbit around the planet in a retrograde direction. Six more small moons lie beyond Triton in irregular orbits – some may be remnants of the original satellite system ejected from closer to the planet.

Triton

Neptune's single giant satellite, Triton has a surface temperature of around -235°C (-391°F), making it one of the coldest places in the solar system. Yet during its 1989 flyby, Voyager 2 revealed unexpected activity on its surface, in the form of geysers belching nitrogen gas and dust into a thin atmosphere. Triton's surface also shows surprising variety, with few craters, a bright 'south polar cap' and much of the western hemisphere covered in pitted 'cantaloupe terrain'.

Triton's current appearance is thought to be linked to its history. It probably originated as an independent ice dwarf world similar to others found in the Kuiper Belt (see page 120), but was captured into orbit around Neptune during a close encounter. Its arrival in an eccentric, retrograde orbit would have played havoc with the original system of Neptunian satellites, while the huge tidal forces it experienced heated its interior and drove widespread geological activity as they pulled its orbit into the perfect circle seen today.

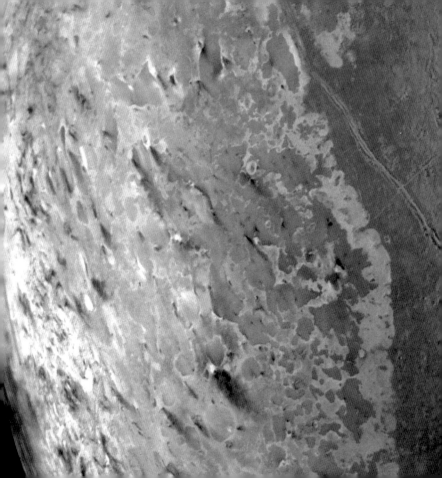

The Kuiper Belt
and Scattered Disc

At distances between 30 and 50 AU from the Sun, the solar system is surrounded by a doughnut-shaped ring known as the Kuiper Belt and a diffuse outer halo called the Scattered Disc. Both are home to small, icy worlds, ranging in size from comets a few hundred metres wide, up to full-blown dwarf planets several thousand kilometres across. They probably formed considerably closer to the Sun and were later ejected into their present orbits by the outward migration of the giant planets early in the solar system's history. The largest known worlds in the Kuiper Belt are the former planet Pluto, and the much more recent discoveries Haumea and Makemake, but there are probably many more bodies of comparable size.

The final phase of planetary migration saw Neptune spiral out to its present position, disrupting some of the bodies already in the Kuiper Belt to form the Scattered Disc. Objects in this region, including the dwarf planet Eris, follow tilted elliptical orbits that extend up to 100 AU from the Sun.

Scattered disc

Orbit of Uranus

Orbit of Neptune

Kuiper Belt

Orbit of Pluto

Pluto

When Pluto was first discovered in 1930, astronomers designated this tiny world as a planet in its own right. It was not until the 1990s that it was confirmed to simply be a large and bright Kuiper Belt Object. Finally, in 2006, Pluto was downgraded to the newly created status of 'dwarf planet', alongside objects such as Ceres and Eris.

Pluto is still the largest object in the Kuiper Belt proper, and despite being just two-thirds the size of Earth's Moon, has a thin atmosphere (which may freeze entirely to the surface when Pluto is at its furthest from the Sun). Its surface is covered in methane and water ice, and has a reddish colour due to the presence of carbon-based chemicals called tholins. It also has a complex system of satellites – the enormous Charon and the far smaller Styx, Nix, Kerberos and Hydra. Charon is more than half the size of Pluto itself, and tidally locked so that both bodies rotate with the same period as Charon's orbit and each keeps one face turned permanently towards the other.

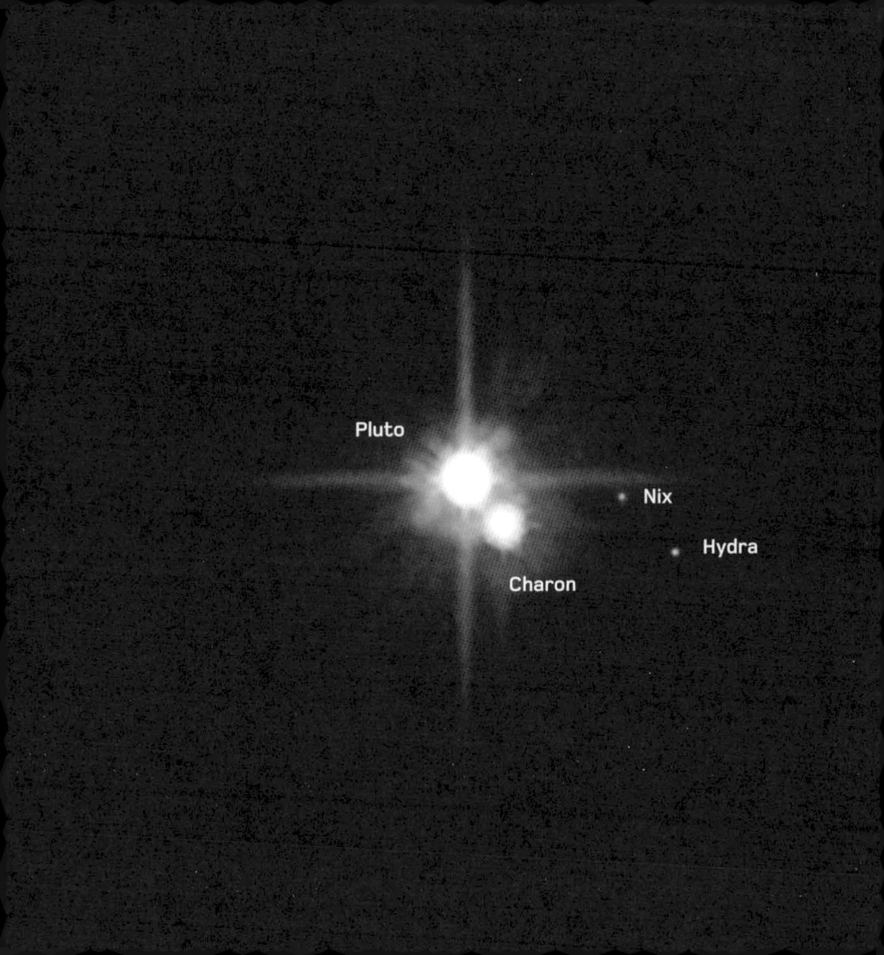
Pluto
Nix
Hydra
Charon

Eris

First identified in 2005 and originally nicknamed 'Xena', Eris is the largest known dwarf planet, and a member of the Scattered Disc that surrounds the Kuiper Belt. Its discovery, and confirmation that it was larger than Pluto, led to the reassessment of Pluto's previous status as a planet.

Eris has a highly elliptical orbit that takes it between 38 and 97 AU from the Sun in the course of some 560 years, inclined at 44 degrees from the plane of the solar system. Its surface appears to be covered in water and methane ice, and is surprisingly bright – in fact, it has the most reflective surface of any known world apart from Saturn's moon Enceladus. Eris is probably a couple of hundred kilometres larger than Pluto, but an estimated 27 per cent more massive – astronomers can estimate its mass because it is accompanied on its long journey around the Sun by a satellite called Dysnomia. Fittingly, for a world whose discovery caused such consternation and discussion, Eris is named after the Greek goddess of discord.

Sedna

Discovered in 2003, Sedna is currently the most distant known object in the solar system, and the first member of a previously hypothetical 'inner Oort Cloud' of ice objects in orbits well beyond the Kuiper Belt. At the time of discovery, it was near perihelion (its closest point to the Sun), a mere 75 AU away. At aphelion, its most distant point, it wanders out to an estimated 937 AU – so distant that a single orbit around the Sun takes around 11,400 years.

Despite Sedna's immense distance, Earth-based telescopes have been able to reveal that it is the reddest object in the solar system – redder even than Mars. Its crust is thought to be a mix of methane and water ice similar to Pluto, with a healthy proportion of carbon-based 'tholin' chemicals providing its red hue. In 2014, astronomers announced the discovery of another object in a similar (though smaller) orbit, lending weight to the idea that the spherical Oort Cloud has an inward extension close to the plane of the solar system.

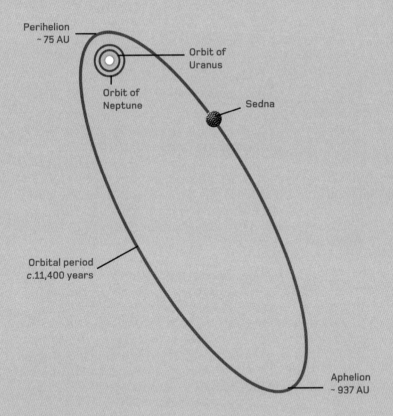

Perihelion
~ 75 AU

Orbit of
Uranus

Orbit of
Neptune

Sedna

Orbital period
*c.*11,400 years

Aphelion
~ 937 AU

The Oort Cloud

A vast spherical halo containing trillions of dormant icy comets surrounds our solar system out to a distance of about 1 light year, and contains the most distant objects still trapped by the Sun's gravitational field. This cloud cannot be observed directly from Earth, but its existence was first inferred by Dutch astronomer Jan Oort in 1932, based on his studies of the orbits of long-period comets, and in particular the locations where they reach their 'aphelia' (most distant points from the Sun).

The comets of the Oort Cloud are exiles from much closer to the Sun: they formed in an icy region beyond the 'frost line' of the early solar system and were later flung out into distant orbits by interactions with the giant planets – and Jupiter in particular. Here they linger on million-year orbits around the Sun, until disruption by rare collisions, or tides raised by close encounters with other nearby stars, send them plunging back towards the inner solar system.

The outer solar system

1 Sun
2 Orbits of planets
3 Kuiper belt
4 Inner Oort cloud
5 Outer Oort cloud

The heliosphere

Despite the presence of the Oort Cloud around 1 light year from the Sun, most astronomers put the boundary of our solar system much nearer than this, in the region where the solar wind falters and dies in the face of the winds from countless other stars. This boundary, the heliopause, coincides roughly with the Kuiper Belt, and defines an egg-shaped region of space known as the heliosphere, blunted in the direction of our solar system's progress through the wider galaxy.

These borderlands are currently being traversed by several interplanetary probes following completion of their primary missions. Missions such as Voyagers 1 and 2, and Pioneers 10 and 11 have sent back data that shows the solar wind slowing down with increasing distance from the Sun until it reaches a point called the 'termination shock', where its speed drops below the speed of sound. Beyond this lies a turbulent region where the solar wind mixes with the interstellar medium, before the heliopause, at which it stops completely.

The edge of the solar system:
1 Sun
2 Solar wind
3 Termination shock
4 Heliopause
5 Bow shock
6 Approaching interstellar medium

The constellations

Throughout astronomy's long history, the constellations have had many meanings. For early stargazers they were symbolic figures in the sky – elements used in fireside story-telling. Later astrologers saw them as patterns through which the Sun and planets moved, either predicting or influencing everything from natural disasters to individual personalities.

In the first century AD, Greek-Egyptian astronomer Ptolemy of Alexandria summarized this ancient view of the sky in a list of 48 constellations, but from the 16th century onwards, European exploration of the southern hemisphere and the invention of the telescope raised many questions. How should the stars of the southern skies be grouped? Where did newly discovered objects fit with the traditional patterns? And what was to be done about the obvious 'gaps' between the old constellations? In the period that followed, many new constellations were promoted – the system was only standardized by the International Astronomical Union in 1930.

Today, the sky is divided into 88 constellations, each defined as an area of the sky. Within these areas, several systems are used to identify individual objects. The brightest stars bear proper names from a variety of sources, but are also assigned Greek letters (so that, at least in theory, the brightest is Alpha, the second brightest Beta, and so on – see page 4). Fainter stars bear a bewildering variety of other designations, ranging from Flamsteed numbers (from catalogues drawn up by British Astronomer Royal John Flamsteed around 1700) to capital-letter designations indicating variable stars. Non-stellar 'deep-sky' objects such as nebulae and galaxies have numbers from their own catalogues, the most comprehensive of which are the Messier list (M) and New General Catalog (NGC).

The brightness of celestial objects, meanwhile, is measured in a comparative system known as magnitude. Originating in classical times, this originally divided all naked-eye stars into six magnitudes, with stars of the first magnitude as the brightest. Today, astronomers use a more rigorous definition, whereby a magnitude difference of 5.0 equals a factor of 100 difference in perceived brightness. This leads to oddities such as negative magnitudes for the brightest stars, but results in a system that can be extended *ad infinitum*.

The constellations by season

Northern circumpolar constellations

These star patterns remain visible to most observers in the northern hemisphere throughout the year.

Northern spring/ southern autumn

These groups are at their highest in the sky after sunset between April and June.

Northern summer/ southern winter

These constellations are best seen in early evening between July and September.

Northern autumn/ southern spring

These groupings are highest in the evening skies between October and December.

Northern winter/ southern summer

These patterns are best located after sunset between January and March.

Southern circumpolar constellations

These patterns are visible to southern-hemisphere stargazers throughout the year.

Ursa Minor

The constellation of Ursa Minor is the northernmost in the sky, and permanently visible from across the northern hemisphere. Its shape resembles the larger Great Bear, Ursa Major (see page 146), with its brightest star, Polaris, marking the tip of the bear's tail.

Polaris, well known as the pole star, currently lies within half a degree of the north celestial pole, the pivot point around which the entire celestial sphere appears to spin as a result of Earth's daily rotation. As a result, it remains almost static in the sky throughout each night. However, Polaris lies in this direction entirely by chance, and the alignment is purely temporary. A slow wobble in the direction of Earth's axial tilt known as precession, causes both the north and south celestial poles to move around the sky in a 25,800-year cycle. Polaris itself is a yellow supergiant roughly 430 light years from Earth. It was once a Cepheid variable star (see page 300), but now shines at a steady magnitude 2.0.

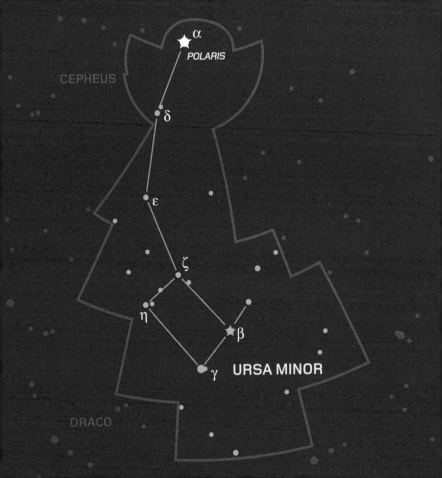

Draco

This large constellation winds its way around the north celestial pole, encircling the polar constellation of Ursa Minor itself. It represents a dragon fought by the mythical demigod Hercules as one of his 12 labours, and the hero himself lies nearby in the sky (see page 152), standing on the dragon's head.

Draco's brightest star is Etamin (Gamma Draconis), an orange giant of magnitude 2.2 some 150 light years from Earth. Etamin and our Sun are approaching in space, and will pass within 28 light years of each other in 1.5 million years time, at which point Etamin may outshine Sirius as the brightest star in the sky. Arrakis (Mu Draconis) is an attractive binary system of yellow-white stars, each with magnitude 5.7, that can be separated with a moderate telescope under high magnification. NGC 6543, meanwhile, is the famous Cat's Eye Nebula, an attractive planetary nebula of magnitude 8.1 about 3,600 light years from Earth.

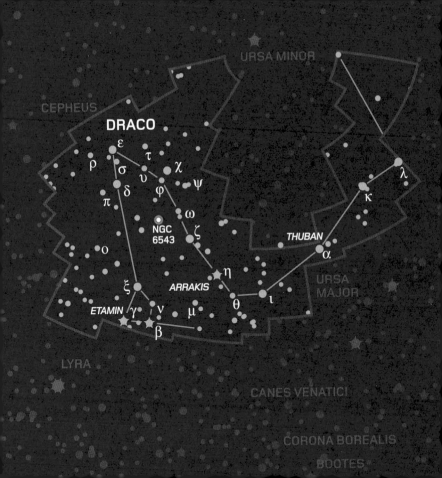

Cepheus and Camelopardalis

These two constellations of far northern skies cover a large area of the sky, but are fairly indistinct. Cepheus is an ancient star pattern representing the ancient King of Ethiopia, husband of Queen Cassiopeia (see page 142), while Camelopardalis was added to the sky in 1613 by Dutch theologian Petrus Plancius.

Cepheus is home to a stretch of the northern Milky Way, and plays host to several renowned variable stars including the unpredictable red supergiant Mu Cephei, also known as the Garnet Star. This is one of the largest and brightest stars known – in our solar system it would extend to the orbit of Jupiter – and emits the same energy as about 350,000 Suns. Another famous variable is Delta Cephei, a yellow supergiant 890 light years away that varies between magnitudes 3.6 and 4.3 in a regular cycle of 5.37 days. Camelopardalis's highlight, meanwhile, is Kemble's Cascade, a beautiful stream of colourful stars leading towards the star cluster NGC 1502.

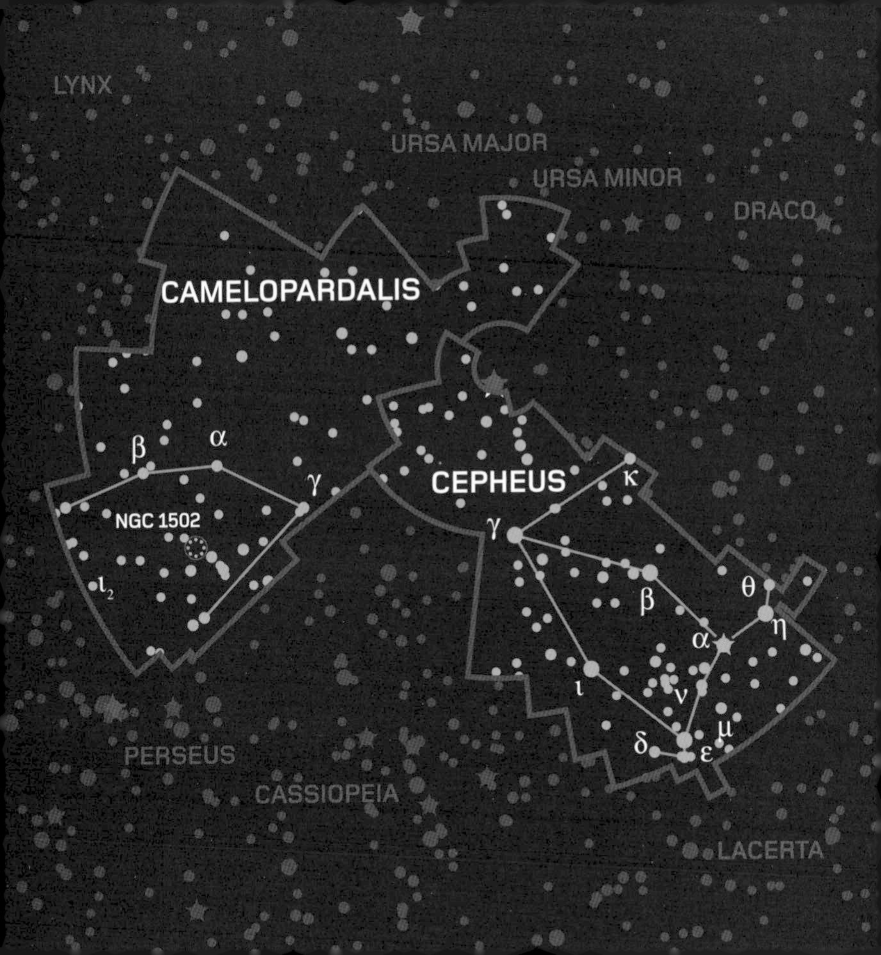
LYNX
URSA MAJOR
URSA MINOR
DRACO
CAMELOPARDALIS
β
α
γ
NGC 1502
ι_2
CEPHEUS
κ
γ
θ
β
η
α
ι
ν
μ
δ
ε
PERSEUS
CASSIOPEIA
LACERTA

Cassiopeia

One of the sky's most distinctive constellations, Cassiopeia's brightest stars form a distinctive W-shaped pattern, circling around the north celestial pole on the opposite side of the sky from the Great Bear, Ursa Major. This ancient grouping represents the vain queen and mother of Andromeda in the Perseus legend, whose boasts of her beauty brought down the wrath of the jealous goddess Hera in the form of the rampaging sea monster Cetus.

Cassiopeia is home to a broad swathe of the northern Milky Way, including beautiful star clusters such as Messier 52 and Messier 103. Its brightest star, Gamma Cassiopeiae, is a fast-spinning blue-white star about 610 light years away, with ten times the mass of the Sun and 40,000 times its luminosity. The slightly fainter Schedar (Alpha Cassiopeiae) is a yellow-orange giant, 230 light years from Earth and about 500 times more luminous than the Sun, while Caph (Beta Cassiopeiae) is a nearby yellow-white giant about 54 light years away.

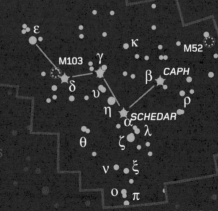

CAMELOPARDALIS

CASSIOPEIA

CEPHEUS

ε

κ

M52

M103

γ

β *CAPH*

δ

υ

ρ

η

α

SCHEDAR

λ

PERSEUS

θ

ζ

ν

ξ

o

π

ANDROMEDA

Auriga and Lynx

The bright stars of Auriga make the ancient constellation of the Charioteer unmistakable in northern winter skies. Its neighbour Lynx is a famously obscure invention of Polish astronomer Hevelius, dating from the 1680s.

Shining at magnitude 0.1 and lying just 42 light years from Earth, Capella (Alpha Aurigae) is a binary consisting of twin yellow-giant stars orbiting too tightly to separate through even the most powerful telescope. Nearby lies a triangle of stars known as 'the Kids' and representing a pair of young goats. Among these, Zeta is an eclipsing binary (see page 302) whose brightness dips from magnitude 3.75 to 4.0 every 972 days. Epsilon Aurigae is a more puzzling system that dips in brightness for a whole year every 27 years, and seems to involve one star being eclipsed by another with a protoplanetary disc (an infant solar system) around it. More immediately appealing for binocular observers is an attractive trio of Milky Way star clusters: Messier 36, M37 and M38.

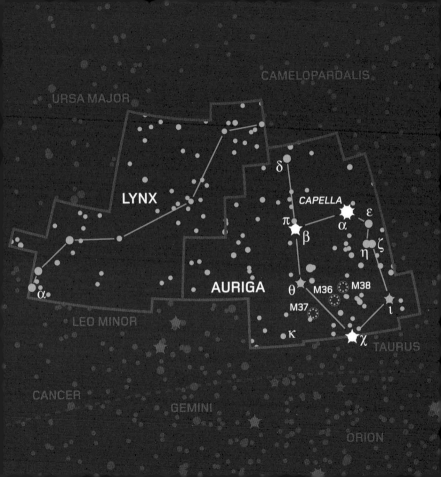

Ursa Major

Perhaps the most famous constellation in the heavens, the Great Bear is a permanent fixture of northern-hemisphere skies throughout the year, and is also widely visible down to mid-southern latitudes. Its seven brightest stars make up the pattern known as the Plough or Big Dipper, but the entire constellation extends well beyond, and is the third largest in the entire sky. The two bright stars on the end of the Dipper's 'pan' are Dubhe and Merak, also known as the Pointers because they point directly towards the pole star Polaris.

Many of Ursa Major's stars originated in the same star cluster about 300 million years ago and are still moving through space together about 80 light years from Earth. For observers, highlights include the famous double star Mizar and Alcor in the bear's tail (in fact, a complex multiple system), the nearby bright spiral galaxy Messier 81 and its irregular companion, the Cigar Galaxy (M82), and the more distant Pinwheel Galaxy (M101), a face-on spiral some 27 million light years from Earth.

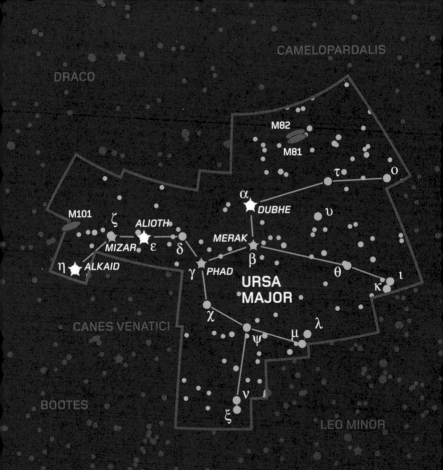

Canes Venatici

Tucked beneath the tail of the Great Bear, the constellation of the Hunting Dogs is small and contains just two stars of any significance, but is nevertheless home to some interesting objects. The pattern was invented by Johannes Hevelius in the late 17th century and represents the dogs of Boötes the herdsman, forever chasing the two bears around the celestial pole.

The brightest star, Alpha Canum Venaticorum (known as Cor Caroli or Charles's Heart in honour of the executed British King Charles I) is a wide binary system some 82 light years from Earth, with two components of magnitudes 2.9 and 5.6 that are easily separated through binoculars. Messier 3, meanwhile, is one of the best globular clusters in the northern sky (see page 308) – just below naked-eye visiblity, appearing as a fuzzy star through binoculars. The constellation's great highlight, however, is the Whirlpool Galaxy (Messier 51) – a face-on spiral some 15 million light years from Earth.

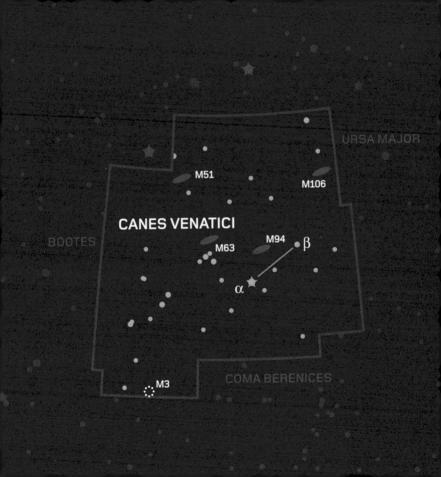

Boötes and Corona Borealis

These two ancient constellations form a long kite shape and a tight arc of stars in northern skies. Boötes represents a herdsman chasing the bears away from his flock, while Corona Borealis is said to be the wedding coronet of Princess Ariadne. Arcturus (Alpha Boötis) is the fourth brightest star in the sky – a red giant of magnitude -0.04, some 36 light years away. Izar (Epsilon Boötis), meanwhile, is a beautiful double star 150 light years away, with orange and blue components separable by small telescopes. And Tau Boötis may shine at an uninspiring magnitude 4.5, but it is home to one of the first planets found orbiting another star, some 51 light years from Earth.

Corona Borealis is home to two famous variables: the 'Blaze Star' T Coronae Borealis (a recurrent nova that brightens from magnitude 11 to 2 every few decades), and the mysterious R Coronae Borealis – a distant yellow supergiant of magnitude 5.8 whose brightness plunges every few years as it obscures itself behind a huge dust cloud of its own making.

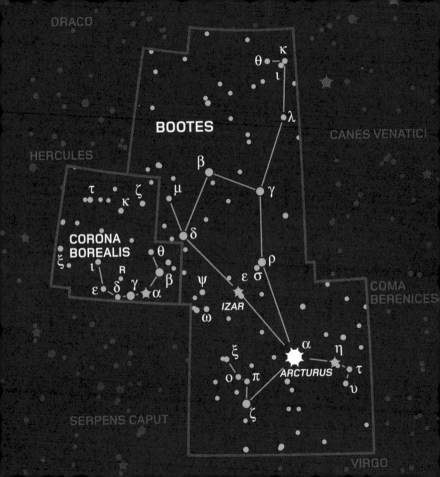

Hercules

This large, straggling constellation represents the Greek demigod famous for his 12 mythological labours. His body is represented by a rough square of stars called the Keystone, from which arms and legs emerge as loose chains of stars. He is usually represented 'upside down', with his foot on the head of the dragon Draco (see page 138).

Rasalgethi (Alpha Herculis) is a well-separated multiple star 380 light years from Earth. Its brightest components are an unstable red giant that varies in brightness between magnitudes 2.8 and 4.0, and a fainter yellow giant of magnitude 5.3 (with an even fainter yellow-dwarf companion of its own). However, the undoubted highlight of Hercules is Messier 13, the finest globular cluster in the northern sky. This huge ball of around 300,000 stars lies about 25,000 light years from Earth and shines on the edge of naked-eye visibility. Binoculars show it as a fuzzy ball, while even small telescopes begin to resolve individual stars around its edges.

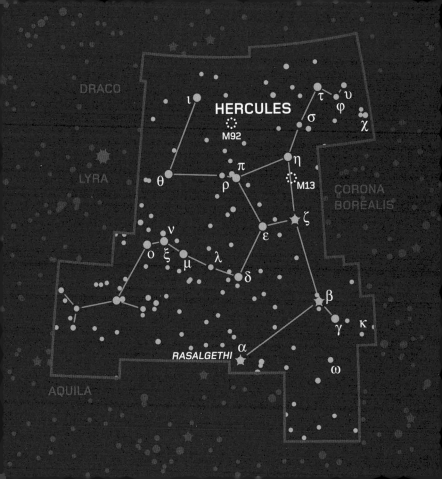

Lyra

The compact but distinct constellation of Lyra represents the ancient stringed instrument played by the Greek hero Orpheus on his journey into the Underworld. Its brilliant star Vega (Alpha Lyrae) shines at magnitude 0.0 and marks one corner of a 'summer triangle' of bright northern stars (with others marked by Deneb in Cygnus and Altair in Aquila).

Vega is a young white star, 25 light years from Earth, 50 times more luminous than the Sun, and still surrounded by a debris disc that may have formed a planetary system. The Ring Nebula (Messier 57), in contrast, marks the location of a star at the other end of its life cycle – a glowing planetary nebula created as a dying Sun-like star 1,100 light years from Earth puffs its outer layers into space. Shining at magnitude 9.5, it is best seen through a small telescope. Epsilon Lyrae, meanwhile, is a famous 'double double' star that reveals two components through binoculars, each of which can be split again through a telescope.

Vulpecula and Sagitta

Two small constellations lie to the south of Cygnus – a tiny but distinctive arrow that has been recognized since ancient times, and a far less well-defined constellation that was originally the 'Fox and Goose', invented by Johannes Hevelius in the late 17th century. Vulpecula's most distinctive features are two deep-sky objects – a coathanger-shaped cluster of faint stars known as Brocchi's Cluster and Messier 27, the brightest planetary nebula in the sky. Visible through binoculars under dark skies, this shell of gas puffed out from a dying star is a quarter the diameter of a Full Moon and lies 1,000 light years from Earth.

Alpha and Beta Sagittae are genuine neighbours in space – twin yellow stars of magnitude 4.4, about 470 light years from Earth. Sagitta's brightest star, however, is Gamma Sagittae, an orange giant at the tip of the arrow, 175 light years away. The constellation's sole Messier object, M71, is a loose globular cluster, 13,000 light years away and just visible in binoculars.

Cygnus

This bright and easily spotted constellation represents a swan flying down the length of the northern Milky Way. To the naked eye its most obvious feature is the Cygnus Rift – a dark 'gap' in the Milky Way created where a nearby dust cloud blocks the light of more distant star clouds.

Deneb (Alpha Cygni) shines at magnitude 1.3, but it is in fact one of the most luminous stars in the sky – 2,600 light years from Earth and 160,000 times brighter than the Sun. Albireo (Beta Cygni) is a celebrated double star – binoculars reveal a beautiful pairing of yellow and blue stars of magnitudes 3.1 and 4.7 respectively, some 385 light years from Earth. The star 61 Cygni is a fainter but closer double – a wide pair of orange stars of magnitudes 5.2 and 6.0 some 10.4 light years away – and was the first star to have its distance accurately determined. The North America Nebula (NGC 7000), meanwhile, is one of the sky's largest emission nebulae, best seen through binoculars or a wide-field telescope.

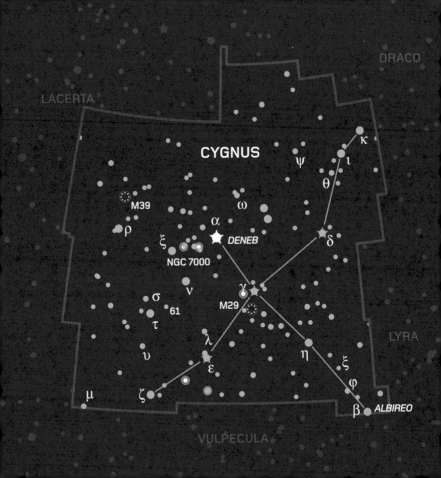

Andromeda and Lacerta

The constellation of Andromeda is marked by two chains of stars rooted together at the northeast corner of the Square of Pegasus. It represents the daughter of King Cepheus and Queen Cassiopeia, doomed to be sacrificed to the monster Cetus before being rescued by the hero Perseus. Lacerta is a 17th-century invention of Johannes Hevelius, with a pattern that resembles a scurrying lizard.

Alpheratz (Alpha Andromedae) is a blue-white magnitude-2.1 star, some 97 light years from Earth. Analysis of its light reveals that it is a binary system, unresolvable through even the most powerful telescope. Gamma Andromedae, meanwhile, is a beautiful multiple with three components separable through telescopes of varying size and magnification. Andromeda's most famous objects, however, are its galaxies – the enormous spiral Messier 31 (the most distant object visible to the naked eye, shining at magnitude 3.4 across 2.9 million light years), and its compact satellites M32 and M110.

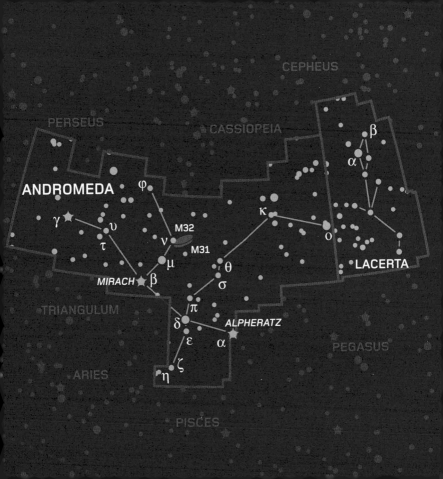

Perseus

This bright but somewhat shapeless northern constellation straddles the Milky Way between Cassiopeia and Auriga. It represents the Greek hero who killled the gorgon Medusa and rescued the princess Andromeda from the sea monster Cetus. Perseus is depicted brandishing Medusa's head in his left hand, and the famous variable star Algol (Beta Persei) marks the gorgon's eye. This was the first eclipsing binary to be identified – a double-star system whose stars orbit each other every 2.87 days, producing periodic eclipses that see its brightness drop from 2.1 to 3.4 for about 10 hours.

Mirphak (Alpha Andromedae) is a yellow supergiant 590 light years from Earth, surrounded by a cluster of massive blue and white stars known as Melotte 20 or the Perseus OB1 association. Midway between Mirphak and neighbouring Cassiopeia lies the famous Double Cluster – a pair of bright open star clusters 7,000 light years away. Designated NGC 869 and NGC 884, they make a beautiful sight in binoculars.

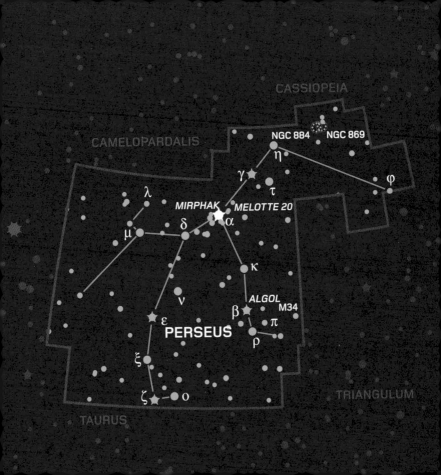

Pisces

This faint constellation consists of two roughly perpendicular trails of stars representing the bodies of two fish, tied together at their brightest star, Alpha Piscium or Alrescha ('the Cord'). In classical myth, the fish represent Venus and Cupid, who transformed themselves in order to escape the ravaging monster Typhon. Alrescha is actually a binary consisting of two white stars of magnitudes 4.2 and 5.2, some 140 light years from Earth. The two components can be separated with a moderate telescope, but the fainter star is probably a spectroscopic binary in its own right. Zeta Piscium, meanwhile, is a fainter but wider binary at about the same distance, with components of magnitudes 5.2 and 6.3.

Messier 74 is a beautiful spiral galaxy with a well-defined 'grand design' structure. Shining at magnitude 10 it is just visible through small telescopes, though much larger instruments are required to show it in detail. It lies at the centre of a small galaxy group some 30 million light years from Earth.

Aries and Triangulum

The indistinct zodiac constellation of Aries represents the Winged Ram of Colchis, whose golden fleece was sought by Jason and the Argonauts. Nearby lies the compact triangle, also recognized by the ancient Greeks, who saw it as the capital letter delta from their alphabet.

Hamal (Alpha Arietis) is a yellow giant some 66 light years from Earth, shining at magnitude 2.0, while Mesartim (Gamma Arietis), is an attractive double with twin white components of magnitude 4.8, some 200 light years away. Lambda Arietis is a less well-balanced double, with a magnitude-4.8 white primary and a yellow companion of magnitude 7.3.

Triangulum's brightest star is a spectroscopic binary of combined magnitude 3.0 about 130 light years away. However, this small constellation's most famous object is the Triangulum Galaxy (Messier 33); a loose face-on spiral 2.7 million light years away, best viewed through binoculars under a dark sky.

Taurus

Unusually among constellations, Taurus bears a striking resemblance to the creature it is supposed to represent, and depictions of this charging celestial bull have been found in cave paintings, perhaps 18,000 years old. The constellation has a wealth of interesting objects, including the V-shaped Hyades star cluster, a group of about 200 stars some 150 light years from Earth. The constellation's brightest star Aldebaran (Alpha Tauri) lies in the foreground of the Hyades as seen from Earth. Lying about 65 light years from Earth, it is an unpredictably variable red giant of average magnitude 0.85.

Another famous star cluster, the Pleiades or Seven Sisters (Messier 45), marks the bull's hunched back – 440 light years away, it contains at least 1,000 young stars and is easily visible to the naked eye. The Crab Nebula (Messier 1) represents the other end of a stellar life cycle – a fuzzy magnitude-8.4 cloud of gas that is the sky's most obvious supernova remnant – the shattered remains of an exploded massive star.

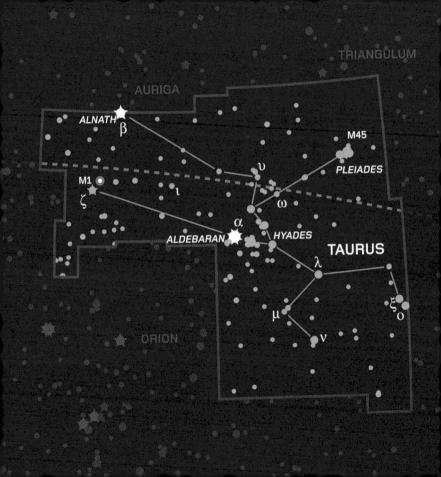

Gemini

The constellation of the Twins is easily identified thanks to its two prominent stars of roughly equal brightness. Castor and Pollux are named after the mythological twin sons of Queen Leda of Sparta who later joined the crew of the Argo on its quest for the Golden Fleece. Fittingly, Castor (designated Alpha Geminorum, although it is actually the constellation's second brightest star) is a fascinating multiple star in its own right. Telescopes reveal twin white stars about 51 light years away with a faint red companion, and analysis of their light shows that each of these stars is itself a double. Their combined magnitude of 1.6, however, is outshone by Pollux (Beta Geminorum) – a single yellow giant some 34 light years away, shining at magnitude 1.1.

The southwestern side of the constellation lies within the Milky Way and is home to some beautiful deep-sky objects, including the open cluster Messier 35 and the compact Eskimo Nebula, a planetary nebula (NGC 2392), 3,000 light years from Earth.

Cancer

The faintest of the 12 zodiac constellations, Cancer is best located by looking midway between its brighter neighbours Gemini and Leo. It represents a crab that attacked the demigod Hercules during his battle with the water serpent Hydra, and was crushed underfoot during the struggle.

Cancer's brightest star is the orange giant Altarf (designated Beta Cancri). About 290 light years from Earth, it shines at magnitude 3.5. Acubens (Alpha Cancri), meanwhile, is a white star of magnitude 4.2, some 175 light years away. Despite a location some distance from the great star clouds of the Milky Way, the constellation is home to two bright star clusters. Messier 67 is an unusually compact and ancient open cluster that has somehow held itself together through its own gravity for about 4 billion years. Nevertheless, at a distance of 3,000 light years, it lies just below naked-eye visibility. Messier 44, in contrast, is a young, bright naked-eye cluster just 580 light years away, also known as the Beehive Cluster.

Leo and Leo Minor

The famous constellation of Leo bears a strong resemblance to a crouching lion, and lies just to the north of the celestial equator. It is one of the most ancient constellations, whereas the small and faint Leo Minor is a much later invention, devised by Johannes Hevelius in the late 1600s.

Leo's brightest star is Regulus (Alpha Leonis), a blue-white star of magnitude 1.4, some 80 light years from Earth, with a magnitude-7.8 companion visible through binoculars. Algieba (Gamma Leonis), meanwhile, is an attractive double whose components – yellow giants of magnitudes 2.0 and 3.2, 170 light years from Earth – can be split through a small telescope. R Leonis is a long-period variable – a red giant that usually lies below naked-eye visibility but peaks at magnitude 4.3 every 312 days or so. Leo also contains two significant groups of galaxies – Messier 95, M96 and M105 beneath the lion's belly, and M65 and M66 in its hind leg. M105 is an elliptical ball of stars, while the others are spiral in structure.

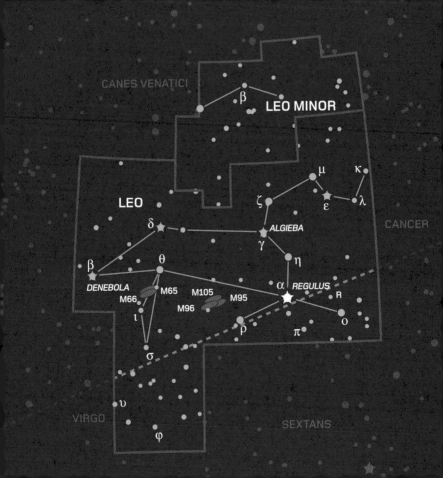

Coma Berenices

Although it contains no bright stars and lacks a prominent shape, Coma Berenices is easily located thanks to its position between Boötes and Leo, and the presence of Melotte 111, a nearby star cluster with at least 20 stars visible to the naked eye under a dark sky. The strands formed by these stars give rise to the Coma's name – it commemorates the locks of Berenice, a historical queen of Egypt in the third century BC. As might be expected, Melotte 111 is a beautiful sight through binoculars or a low-powered telescope. Messier 53, meanwhile, is a magnitude-8.3 globular cluster 56,000 light years away.

Coma also plays host to many galaxies. Several are related to the Virgo Cluster (see page 178), but Coma has a more distant cluster of its own, some 300 million light years away and containing more than 1,000 galaxies. Most are visible only through powerful telescopes, but the Black Eye Galaxy (Messier 64), a spiral with a prominent dust lane just 25 million light years from Earth, is visible through smaller instruments.

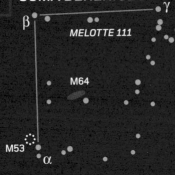

CANES VENATICI

BOOTES

COMA BERENICES

γ

β

MELOTTE 111

M64

M53

α

VIRGO

Virgo

Despite no obvious resemblance to any particular figure, the constellation of Virgo has been associated with a celestial maiden (often a harvest goddess) since the most ancient times. Its brightest star, Spica (Alpha Virginis), has a name that means 'the ear of wheat', and is a spectroscopic binary system with two brilliant blue stars that orbit each other in just four days. Seen over a distance of 260 light years, they shine together at magnitude 1.0. Porrima (Gamma Virginis), meanwhile, is a much wider binary – its twin yellow-white components, 32 light years from Earth, orbit each other in 169 years and are easily separable through small telescopes.

Virgo's greatest fame comes from its concentration of galaxies – it marks the centre of the nearest major galaxy cluster (the Virgo Cluster), around 55 million light years from Earth, and containing at least 1,300 galaxies. The most prominent of these are the elliptical Messier 49 and the giant elliptical Messier 87, marking the very centre of the cluster.

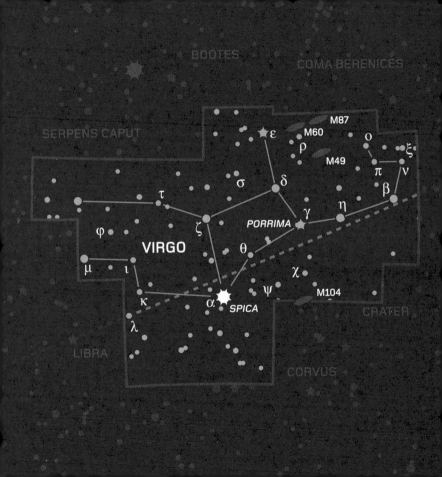

Libra

The only constellation in the zodiac to represent an object rather than an animal, Libra was once relegated to a subsidiary role as the claws of neigbouring Scorpius. Since Roman times, however, it has been seen as the scales of justice, held aloft by the figure of Virgo, its western neighbour.

Libra's brightest stars still retain names that recall the more ancient association. Alpha Librae, or Zubenelgenubi ('Southern Claw'), is a double that can easily be separated with binoculars or even the naked eye, consisting of a blue-white giant of magnitude 2.8 and its white magnitude-5.2 companion some 70 light years from Earth. Beta Librae, or Zubenelschamali ('Northern Claw'), is actually slightly brighter at magnitude 2.6, and is a rare example of a star that appears green to some observers. It also shows periodic dips in brightness that may be due to eclipses by an unseen companion in an unconfirmed eclipsing binary system. Zubenelschamali lies about 160 light years away.

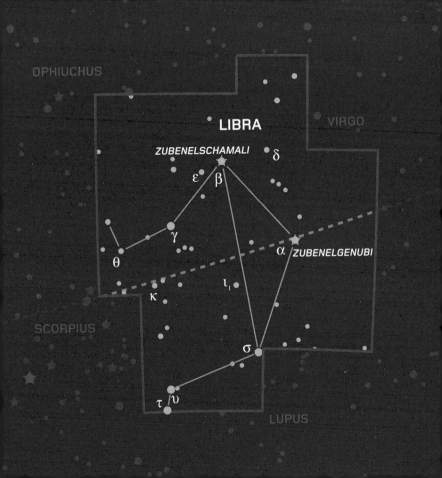

Serpens

The only constellation in the sky that is split into two isolated sections, Serpens represents a snake writhing in the grasp of the giant figure of Ophiuchus (see page 184): this separates it into Serpens Caput; the serpent's head, and Serpens Cauda; its tail. The constellation's brightest star, Unukalhai (Alpha Serpentis), marks the creature's neck and is an orange giant of magnitude 2.6, some 73 light years from Earth.

Messier 5 in Serpens Caput is the brightest of several globular clusters in this region of the sky, just on the border of naked-eye visibility at magnitude 5.6. Despite its distance of 24,500 light years, this ball of stars 170 light years wide is an impressive sight through binoculars or a small telescope. Messier 16 in Serpens Cauda consists of an impressive open cluster surrounded by the huge Eagle Nebula. The Eagle is home to the famous 'pillars of creation' – star-forming regions photographed in detail by the Hubble Space Telescope.

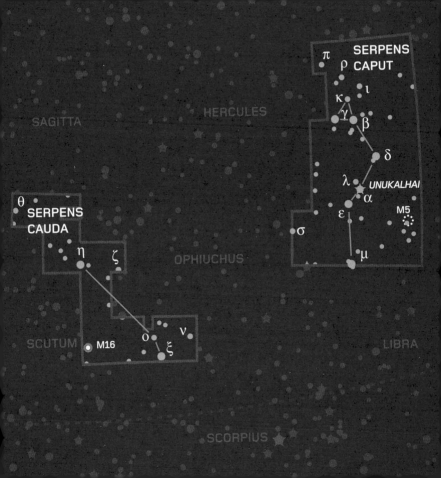

Ophiuchus

This large constellation has been seen as a giant wrestling with the writhing snake Serpens (see page 182) since ancient times, and today is usually associated with Asclepius, the Greek god of healing.

Sadly, despite its size, Ophiuchus' stars are faint and its pattern amorphous. Its brightest star, Rasalhague (Alpha Ophiuchi), marks the figure's head, and is a white giant of magnitude 2.1, 50 light years from Earth. Rho Ophiuchi is a faint but attractive double star. Binoculars show two faint, wide companions of the magnitude 5.0 primary star, still embedded in the nebulosity from which all three formed. A small telescope will reveal Rho's second component – a closer neighbour of magnitude 5.9. Close to Beta Ophiuchi lies Barnard's Star, a magnitude-9.5 red dwarf visible through small telescopes. It is significant because it is the fourth closest star to the Sun (after the Alpha Centauri system – see page 206), and the fastest-moving star in the sky.

Aquila and Scutum

Long-identified as an eagle in flight, Aquila is a small constellation that draws attention to itself through its brightest star. Altair (Alpha Aquilae) marks the southern tip of the northern hemisphere's 'summer triangle' – a large pattern of bright stars completed by Vega in Lyra and Deneb in Cygnus. Altair is one of the nearest bright stars – just 17 light years from Earth and shining at magnitude 0.8 – and is closely flanked by Alshain and Tarazed (Beta and Gamma Aquilae respectively). Alshain is the fainter of this pair, a yellow star of magnitude 3.7, some 45 light years from Earth, while magnitude-2.7 Gamma is an orange giant about 395 light years away.

Nearby Scutum, the Shield, is a small, kite-shaped constellation invented by Polish astronomer Johannes Hevelius and originally named Scutum Sobieski in honour of his patron, King John III Sobieski. It is crossed by the Milky Way and is home to the notably bright Scutum Star Cloud, and an attractive open cluster known as the Wild Duck (Messier 11).

Delphinus and Equuleus

The small but unmistakable constellation of the Dolphin lies to the west of Pegasus, and is said to represent the animals that saved the ancient Greek musician Arion from drowning at sea. The foal Equuleus is a far less obvious pattern, but it, too, appeared in Ptolemy's original list of 48 constellations, representing Celeris, the swift-footed offspring (or sometimes brother) of Pegasus. Curiously, the Alpha stars of each constellation (Sualocin in Delphinus and Kitalpha in Equuleus) both have about the same brightness (magnitudes 3.8 and 3.9 respectively) and both lie about 190 light years from Earth. However, Sualocin is blue-white in colour while Kitalpha is a yellow giant.

Epsilon Equulei is an apparent triple star that actually combines a genuine binary system of magnitude 5.4 (200 light years away and only separable with a large telescope), with a magnitude-7.4 interloper that just happens to lie in the same direction, but is 75 light years closer to Earth.

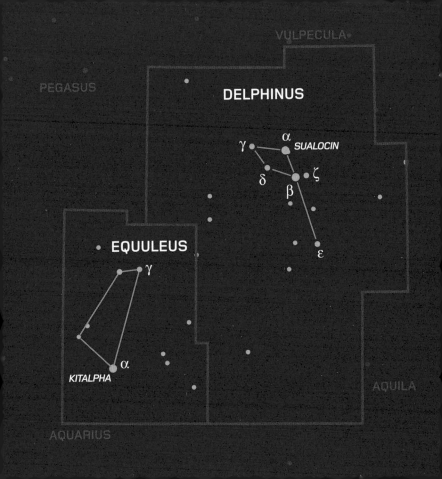

Pegasus

Although Pegasus is one of the largest constellations in the sky, it actually covers a fairly empty area with few naked-eye stars. Nevertheless, it is easily identified because its four brightest stars (one shared with neighbouring Andromeda) form an obvious pattern – the Square of Pegasus. In Greek myth, Pegasus was a winged horse born from the blood of the gorgon Medusa and later ridden by the hero Bellerophon.

Markab (Alpha Pegasi) is a blue-white star 140 light years away, shining at a steady magnitude 2.5, which normally makes it the constellation's brightest star. Scheat (Beta Pegasi) is a red giant that varies unpredictably in brightness – 200 light years from Earth, it usually averages magnitude 2.7, but can occasionally outshine Markab, or dip fainter than magnitude-2.8 Gamma Pegasi. Pegasus' deep-sky highlight is Messier 15, a fine globular cluster of magnitude 6.2. Easily spotted in binoculars as an elliptical ball of light, it lies 34,000 light years from Earth and contains about 700,000 stars.

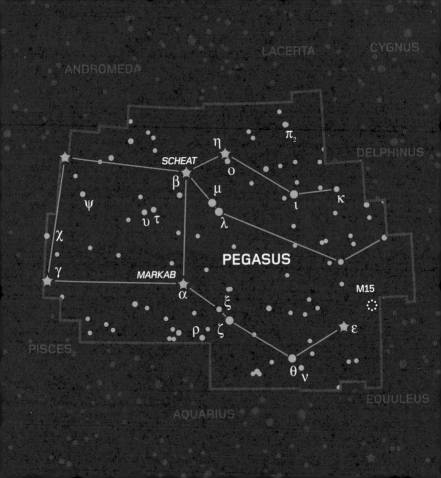

Aquarius

Despite its lack of an obvious shape, this constellation has been associated with a figure pouring water from a jug since at least 2000 BC and today is usually identified as Ganymede, cup bearer to the gods of Mount Olympus. The Y-shaped group of stars around Zeta Aquarii is the Water Jug. Zeta itself is a beautiful binary star about 103 light years from Earth, with near-twin white components of magnitudes 4.3 and 4.5, easily separated through a small telescope. Sadalmelik and Sadalsuud (Alpha and Beta Aquarii) are yellow supergiants 760 and 610 light years from Earth, both shining at magnitude 2.9. Both these stars and Epsilon Pegasi, are following similar trajectories out of the plane of our galaxy, from a probable origin in the same open star cluster.

Messier 2 is a magnitude-6.5 globular cluster, easily visible through binoculars, while NGC 7009, the Saturn Nebula, is a compact magnitude-8.0 planetary nebula some 2,400 light years away, and a good target for small telescopes.

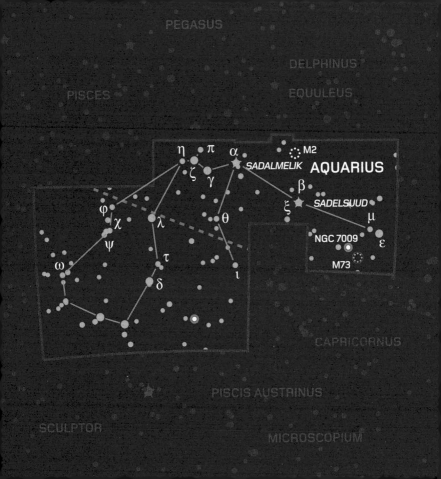

Cetus

Although Cetus' name is more properly translated as 'the Whale', since ancient times it has been asssociated with the fearsome sea monster Typhon, sent to ravage the lands of King Cepheus and Queen Cassiopeia in the Perseus legend. The constellation straddles the celestial equator, and can alter its appearance substantially depending on the behaviour of its brightest star, Mira (Omicron Ceti). Mira, whose name means 'wonderful', was the first variable star to be recognized, as early as 1596, and is the prototype for a class known as the long-period variables. Its brightness varies between a peak around magnitude 3 and a minimum around magnitude 10, as it pulsates in size (see page 300) over a period of about 332 days.

Tau Ceti is a relatively Sun-like star of magnitude 3.5, just 11.9 light years from Earth, surrounded by a disc of planet-forming debris and possibly orbited by up to five planets. UV Ceti is an even closer red-dwarf binary, just 8.7 light years away but, at magnitude 12, only visible through large telescopes.

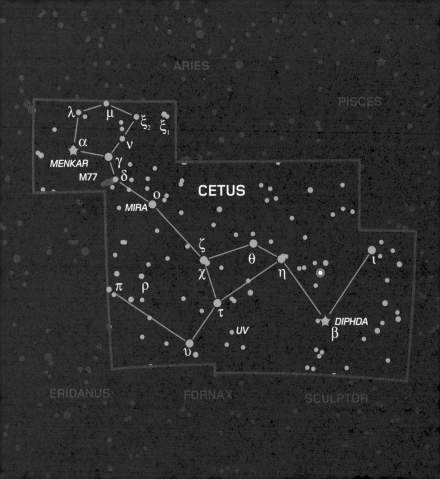

Orion

One of the sky's most famous and recognizable constellations, the great hunter Orion spans the celestial equator as he faces the charging bull Taurus. Two of the brightest stars in the sky, blood-red Betelgeuse (Alpha Orionis) and blue-white Rigel (Beta Orionis) mark Orion's shoulder and knee respectively. Rigel is the brighter of the two – a blue supergiant about 860 light years from Earth, shining at magnitude 0.1 with a luminosity 66,000 times that of the Sun. Betelgeuse is a red supergiant that varies between magnitudes 0.3 and 1.2, about 640 light years away.

Orion marks the centre of one of the closest star-forming regions to Earth, about 1,400 light years distant. Faint nebulosity fills the constellation, but is at its brightest in the Great Orion Nebula (Messier 42). Visible to the unaided eye as a 'fuzzy star' in Orion's sword, this complex cloud of gas and stars runs south from the line of three bright stars forming the figure's belt, and includes several attractive multiples.

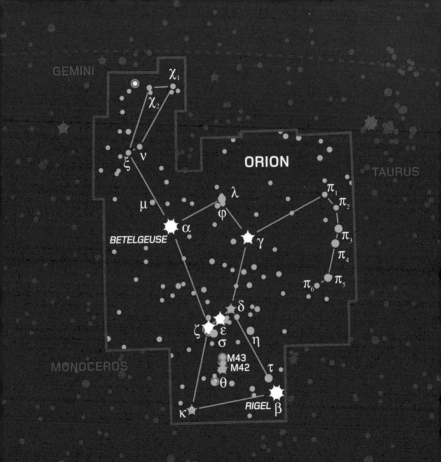

Monoceros and Canis Minor

Directly to the east of Orion lie the faint stars of the Unicorn (a large W-shape that may have originated among Arabic or Persian astronomers). Nearby lies the compact Lesser Dog, which at least draws attention to itself through the presence of magnitude-0.3 Procyon (Alpha Canis Minoris) one of the brightest stars in the sky. Procyon is a nearby white star 1.4 times the mass of the Sun and seven times more luminous, just 11.4 light years away. Large telescopes can reveal its faint white-dwarf companion (see page 318).

Monoceros is home to several interesting star-forming regions. The Christmas Tree Cluster (NGC 2264) is an open cluster of combined magnitude 3.9, some 2,400 light years from Earth. Its stars illuminate surrounding gas clouds, which in turn silhouette the Cone Nebula – a dark tower of dust within which star formation is currently underway. NGC 2244 is another naked-eye cluster, some 5,500 light years away, surrounded by the beautiful but elusive Rosette Nebula.

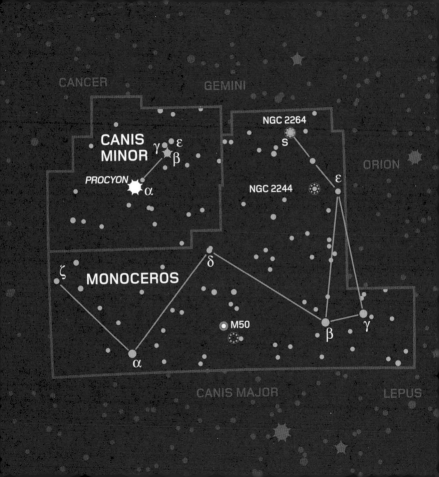

Canis Major

The larger of Orion's two hunting dogs is easily found thanks to the presence of Sirius, the brightest star in Earth's skies at magnitude -1.46. Sirius owes its brightness to its proximity to Earth – it is a fairly normal white star, 25 times more luminous than the Sun but just 8.6 light years away. It is also a binary, though its faint white-dwarf companion (see page 318) is usually hidden in the glare of the brighter star. Wezen (Delta Canis Majoris) offers a marked contrast to Sirius – though apparently fainter at magnitude 1.8, it is in reality a yellow supergiant, 1,800 light years away and about 2,000 times more luminous than Sirius.

Messier 41 is an open cluster – a telescope shows it as a field of about 100 stars dominated by ageing red and orange giants, already near the end of their lives about 190 million years after the cluster's formation. The Tau Canis Majoris Cluster (NGC 2362) is much younger, and still dominated by massive but short-lived blue stars in the prime of their lives.

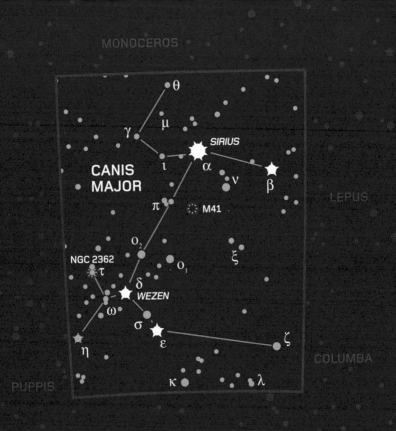

Hydra

The largest and longest constellation in the sky, Hydra the water snake runs along the celestial equator to the south of better-known patterns, such as Leo and Virgo. Ancient in origin, it supposedly represents a snake flung into the sky by the god Apollo, along with the cup, Crater, and a crow, Corvus.

Magnitude-2.0 Alphard (Alpha Hydrae) is a red giant about 180 light years away from Earth, and the constellation's brightest star by some way – its name means 'the solitary one'. East of Hydra's head, near the border with Monoceros, lies the open star cluster Messier 48 – just visible to the naked eye, binoculars will reveal it as a triangular cloud of several dozen stars, about 1,500 light years away. Messier 83 in Hydra's

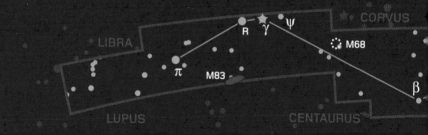

tail, meanwhile, is the Southern Pinwheel Galaxy, a face-on spiral of magnitude 7.6, just 15 million light years away but best seen through larger instruments. NGC 3242, a planetary nebula known as the 'Ghost of Jupiter', has a similar magnitude but is more compact, making it an easier target for small telescopes.

Crater, Corvus and Sextans

Three small constellations run to the south of Hydra – the Cup and the Crow are ancient in origin, while the Sextant was introduced by Polish astronomer Johannes Hevelius in 1687. In mythology, Corvus was a servant of Apollo, sent to fill his cup from a well. When the crow dawdled on the way, and then attempted to blame his tardiness on a snake blocking the well, the god became angry and hurled all three into the sky.

Algorab (Delta Corvi) is an unusual binary star about 88 light years from Earth. Its brighter component is a yellow-white star of magnitude 3.0, but a small telescope will reveal a wide companion of magnitude 8.5 that appears to have a purplish colour – a trick of the eye caused by contrast. NGC 4038 and 4039 are the famous colliding Antennae Galaxies – barely visible through small instruments but imaged in spectacular detail by the Hubble Space Telescope. NGC 3115, meanwhile, is the Spindle Galaxy, an edge-on lenticular galaxy (see page 360) about 13 million light years away.

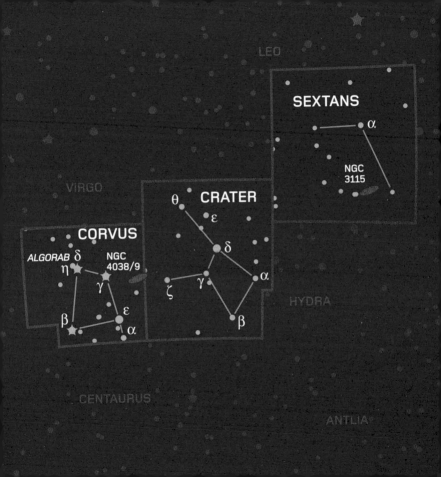

Centaurus

One of two centaurs found among the constellations (the other being Sagittarius, the Archer), Centaurus occupies a large area of the southern sky between Hydra and Crux, the Southern Cross. The famous Rigil Kentaurus (Alpha Centauri) is the closest star system to our own and the third brightest star in the entire sky – a small telescope will split it to reveal a yellow, Sun-like star of magnitude 0.0 with an orange companion of magnitude 1.4. A third component, the red dwarf Proxima Centauri, is the closest star of all. A mere 4.36 light years from Earth, but shining at a feeble magnitude 11, it is hard to detect through small telescopes.

Omega Centauri (NGC 5139), is the most impressive globular cluster in the sky – a ball of a million stars orbiting the Milky Way at a distance of 16,000 light years and easily visible with the naked eye. NGC 5128 is an elliptical galaxy of magnitude 7.0, 15 million light years away. It is crossed by a dark dust lane that shows where it has swallowed up a smaller spiral galaxy.

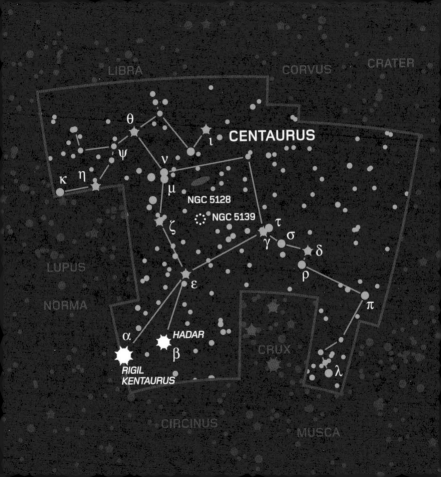

Lupus

Fairly bright but also rather shapeless, this constellation represents a beast attacking nearby Centaurus, and is often shown impaled on the centaur's lance. Although ancient in origin, it only became associated with a wolf in medieval times, and was previously known simply as Bestia or Fera, meaning an unspecified wild beast.

Alpha Lupi (Kakkab) is a blue giant 550 light years from Earth, shining at magnitude 2.3, and a member of the 'Scorpius-Centaurus Association'. This group of blue and white heavyweight stars spreads across several constellations: it represents the brightest members of an open cluster formed about 20 million years ago that has since scattered across nearby space. Mu Lupi, meanwhile, is an attractive multiple star of magnitude 4.3, some 290 light years from Earth. Small telescopes reveal a companion of magnitude 7.2, while larger instruments split the primary blue-white star into two near-twin components, each 70 times more luminous than the Sun.

Scorpius

This bright constellation is one of the most ancient in the heavens, though the scorpion's extended pincers have now been appropriated to form the scales of Libra. The scorpion's body curves its way from a prominent head marked by brilliant red Antares (Alpha Scorpii), to the sting of bright blue Shaula (Lambda Scorpii). Antares is one of the most extreme stars known – a cool but luminous red supergiant large enough to engulf our solar system almost to the orbit of Jupiter. It pulsates erratically around an average magnitude 1.0, and has a magnitude 5.5 companion that is usually lost in its glare.

Acrab (Beta Scorpii) is a more rewarding group – a pair of blue stars of magnitudes 2.6 and 4.9 easily split through a small telescope, with a fainter star of magnitude 10 nearby – while Jabbah (Nu Scorpii) is a 'double double' with four components. Messier 7 is a naked-eye open cluster 800 light years from Earth, while Messier 4 is a globular cluster 7,200 light years away on the edge of naked-eye visibility at magnitude 5.6.

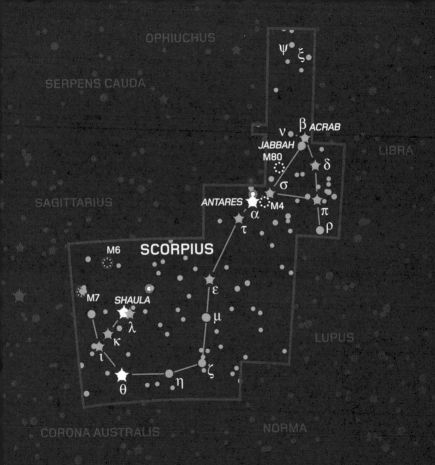

Sagittarius

Usually represented as a centaur (half-human, half-horse), armed with bow and arrow, to modern eyes Sagittarius looks more like a teapot with bright Milky Way star clouds rising like steam from its spout. Sagittarius lies in one of the richest parts of the sky, towards the centre of our galaxy itself (see page 346), and is home to several bright nebulae, largest of which is the Lagoon Nebula (Messier 8), a magnitude-6.0 cloud of glowing gas with an associated naked-eye star cluster. Nearby lie the beautiful Trifid Nebula (M20), separated into three parts by lanes of light-absorbing dust, and the delicate Swan or Omega Nebula (M17). Messier 22, meanwhile, is the brightest of several globular clusters at magnitude 5.1.

The constellation's brightest star is Kaus Australis (Epsilon Sagittarii), a white giant with a Sun-like companion star, shining at magnitude 1.8 at a distance of 145 light years. Arkab (Beta Sagittarii), meanwhile, is a line-of-sight double, whose components lie 380 and 140 light years away.

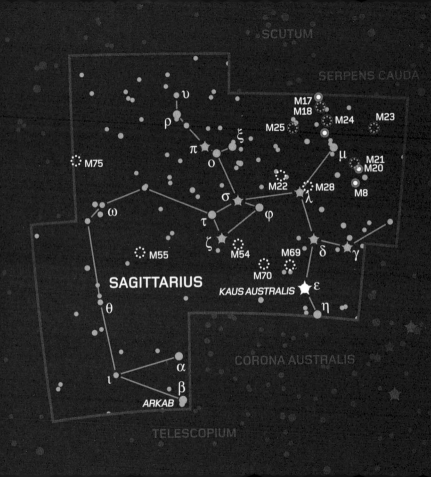

Capricornus

Its stars may be faint and its pattern obscure, but archaeological evidence points to the curious constellation of the Sea-goat being one of the most ancient star patterns of all. This hybrid figure, with the body of a goat and the tail of a fish, has been found in Babylonian and Assyrian art dating back at least 4,000 years.

Algiedi (Alpha Capricorni) is an intriguing combination of line-of-sight double and physical binary stars. Most people can spot its two major components with the naked eye – yellow stars of magnitudes 3.6 and 4.2, 108 and 635 light years from Earth – but small telescopes show that each star has a faint companion of its own. Dabih (Beta Capricorni), in contrast, is a genuine quadruple star, 330 light years distant, but resolving more than two components requires a large telescope. Messier 30 is a globular cluster with overall magnitude 7.2 some 26,000 light years away and best seen through a small- or medium-sized instrument.

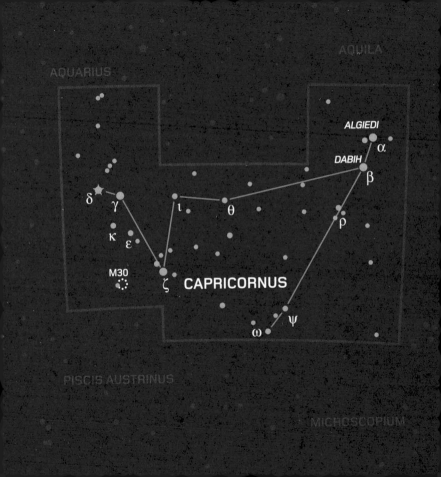

Piscis Austrinus
and Microscopium

The brilliant Fomalhaut draws the eye immediately to the otherwise faint pattern of the Southern Fish, and probably explains why this obscure group of stars was recognized in ancient times. Its neighbour Microscopium, in contrast, was picked out much later – an invention of French astronomer Nicolas Louis de Lacaille in the 1750s. Shining at magnitude 1.2, Fomalhaut (Alpha Piscis Austrini) is a young white star just 25 light years from Earth. Infrared studies and images from the Hubble Space Telescope have shown that it is surrounded by a doughnut-shaped ring of dusty, planet-forming material. Magnitude-4.3 Beta Piscis Austrini is a white star about 148 light years away, with a magnitude-7.7 binary companion that can be spotted with a small telescope.

AU Microscopii is a red-dwarf star 32 light years away. Normally languishing at an obscure magnitude 8.7, it occasionally experiences huge stellar flares on its surface, which cause it to brighten significantly.

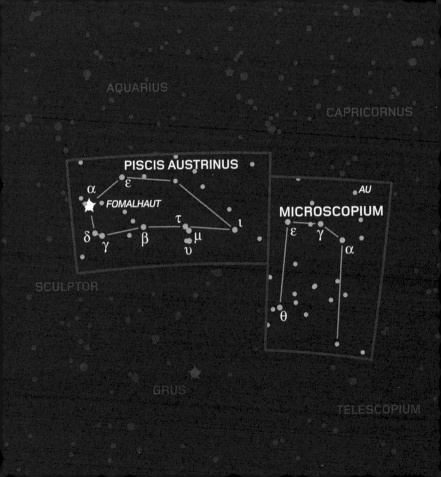

Sculptor and Fornax

These two obscure constellations to the south of Cetus are inventions of the 18th-century French astronomer Nicolas Louis de Lacaille, and bear no resemblance to the objects they supposedly represent. The stars are faint and mostly unremarkable, and the region's chief attractions lie further afield. Sculptor is home to the nearest galaxy group beyond our own Local Group (see page 352), while Fornax plays host to a more distant, but impressive, cluster of several dozen galaxies.

Sculptor's brightest galaxy is the Silver Coin (NGC 253) – an edge-on spiral galaxy of magnitude 7.1, roughly 10 million light years from Earth, and easily spotted through binoculars or a small telescope. Nearby NGC 55 is an irregular galaxy with faint traces of spiral structure – closer to us at about 7 million light years, but fainter at magnitude 8.8. The Fornax Cluster's brightest member is NGC 1316, an unusual lenticular galaxy (see page 360) 70 million light years away and shining at magnitude 9.4.

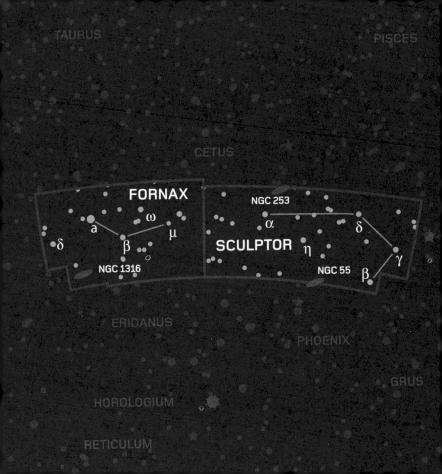

Eridanus

The great celestial river winds its way from the foot of Orion towards far southern skies. Its stars are mostly faint, and its path hard to trace, but the magnitude-2.8 star Cursa (Beta Eridani) lies close to its northern end, and its southern end is marked by Achernar (Alpha Eridani), one of the sky's brightest stars at magnitude 0.5.

Omicron Eridani is one of the constellation's most intriguing stars – a closely spaced line-of-sight double whose fainter component (magnitude-4.4 Omicron-2) is a multiple in its own right. A small telescope will show a magnitude-9.5 companion – the burnt-out remains of a Sun-like star some 16.5 light years from Earth. This is the most easily seen white dwarf in our skies. A larger instrument will reveal another component, a red-dwarf companion of magnitude 11.2. Epsilon Eridani, meanwhile, is one of the closest Sun-like stars to our own. Shining at magnitude 3.4, it lies about 10.5 light years from Earth and is orbited by at least one planet.

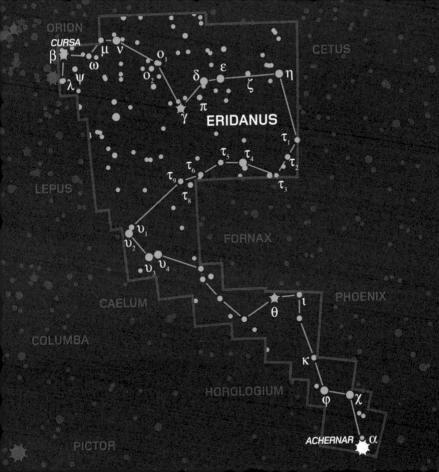

Caelum and Horologium

Three obscure constellations run along the southeastern border of the celestial river Eridanus, each invented by French astronomer Nicolas Louis de Lacaille during the 1750s. They bear little resemblance to the Sculptor's Chisel and Pendulum Clock they are supposed to represent.

Horologium has two deep-sky objects of note – NGC 1512 is a fine barred spiral of magnitude 11.1, 30 million light years away and best seen through a telescope of moderate size. NGC 1261, meanwhile, is a magnitude-8.4 globular cluster at a distance of 54,000 light years. The constellation's most interesting star is R Horologii, a long-period pulsating variable that usually shines at magnitude 14 but peaks every 407 days at magnitude 4.0. Alpha Caeli is a binary system, with a brighter star of magnitude 4.4 accompanied by a red dwarf, normally of magnitude 9.8 but prone to bright flares. Gamma Caeli is another binary – an orange giant of magnitude 4.6 some 185 light years from Earth with a companion of magnitude 8.1.

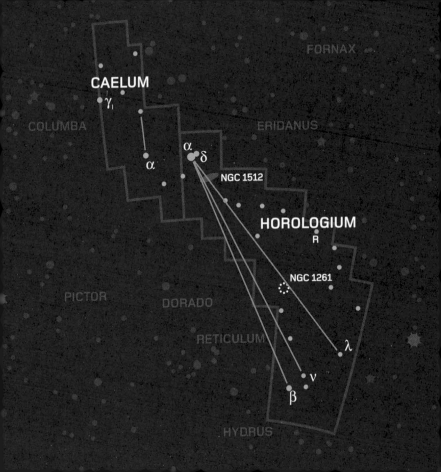

FORNAX

CAELUM

γ₁

COLUMBA

ERIDANUS

α

α δ

NGC 1512

HOROLOGIUM

R

NGC 1261

PICTOR

DORADO

RETICULUM

λ

ν

β

HYDRUS

Lepus and Columba

South of Orion lie two constellations of moderate brightness. Lepus is an ancient Greek star pattern, representing a hare crouching at the hunter's feet. Columba probably originated with Dutch navigators Pieter Dirkszoon Keyser and Frederick de Houtman in the late 16th century.

Arneb (Alpha Leporis) is a rare white supergiant – a star about 30,000 times more luminous than the Sun and 2,200 light years away, shining at magnitude 2.6. R Leporis ('Hind's Crimson Star') is one of the coolest and reddest stars in the sky – a red giant 820 light years away, pulsating in brightness between magnitudes 5.5 and 11.7 over a 430-day period. Phact (Alpha Columbae) is a blue-white star of similar brightness to Arneb, but considerably closer to our solar system at a distance of 530 light years. Messier 79 is a globular cluster of magnitude 7.7, roughly 42,000 light years from Earth in Lepus – its path through the sky suggests that it may have been pulled into orbit around the Milky Way in the relatively recent past.

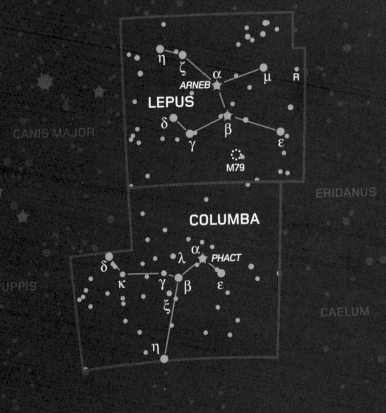

Puppis

This large and bright constellation is the northernmost of three (along with Carina and Vela) that together form the great ship Argo. It represents the stern of the ship used by Jason and his crew on their quest for the Golden Fleece. The constellations were once treated as a single whole, and therefore the Greek letters indicating their bright stars are shared between them – hence Naos, the constellation's brightest star, is designated Zeta Puppis. One of the hottest and most luminous stars known, it has a surface temperature of 42,000°C (75,600°F) and a total energy output 750,000 times greater than the Sun's (largely in the ultraviolet). Across a distance of 1,400 light years, it shines at magnitude 2.2.

Puppis's location along a bright stretch of the Milky Way makes it rich in open star clusters, the most impressive of which is Messier 46. Seen through a small telescope, it reveals 150 individual stars shining at a combined magnitude of 6.0 across 5,400 light years.

Antlia and Pyxis

Sandwiched between the serpentine body of Hydra to the north and the brighter stars of Vela to the south, these two obscure constellations are inventions of the French astronomer Nicolas Louis de Lacaille, who filled in most of the remaining gaps in southern-hemisphere skies during the 1750s. Sadly, however, they contain few objects of interest.

Binoculars will split magnitude-5.7 Zeta Antliae to reveal a closely aligned pair of near-identical white stars at more or less the same distance from Earth (372 and 374 light years). Despite their proximity in space, they are not thought to be bound together by gravity – but a small telescope will show that the western component is a binary star in its own right. T Pyxidis, meanwhile, is a recurrent nova system (see page 336) some 6,000 light years away that normally languishes beyond the reach of amateur telescopes at magnitude 13.8. It occasionally surges to binocular or even naked-eye visibility during eruptions that seem to occur every 20–30 years.

Vela

The sail of the great celestial ship Argo is formed from a rough octagon of bright stars, crossed by a broad swathe of the southern Milky Way. Its brightest star is Regor (Gamma Velorum) – a complex multiple system with a primary star of magnitude 1.8, some 1,200 light years from Earth. Telescopes will reveal further components of magnitudes 4.3, 8.5 and 9.4, the faintest of which is a binary in its own right. The primary, meanwhile, is another 'spectroscopic' binary, consisting of a hot blue star orbited by an even hotter 'Wolf-Rayet' star – a former monster star that has blown away much of its material to leave its inner layers exposed.

Delta Velorum is another complex multiple star, 80 light years away, whose magnitude-2.0 brightest component is an eclipsing binary that drops in brightness by 0.4 magnitudes every 45 days. IC 2391, meanwhile, is a beautiful open cluster surrounding the magnitude-3.2 star Omicron Velorum, at a distance of some 580 light years away.

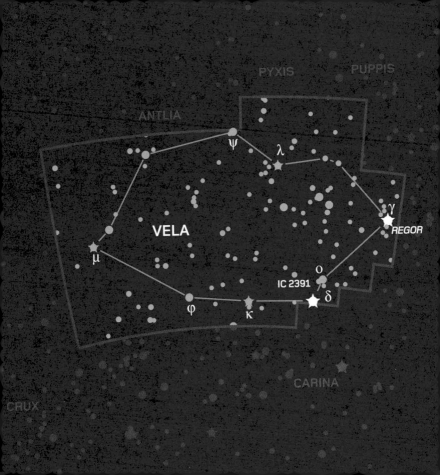

Carina

The brightest section of the great ship Argo is its keel, easily identified by the presence of Canopus (Alpha Carinae), the second brightest star in the sky. Shining at magnitude 0.7, it is a yellow-white supergiant 315 light years from Earth and roughly 15,000 times more luminous than the Sun.

Carina is also home to the sky's most impressive star-forming nebula. Catalogued as NGC 3372, it covers an area of the Milky Way about four times the diameter of the Full Moon, with an overall magnitude of 1.0. The nebula is large enough to have several distinct features within it, including several bright knots of stars and a dark dust lane called the Keyhole Nebula. Most impressive of all is the double-lobed Homunculus Nebula, with the unstable Eta Carinae at its heart — an extreme double star — one of whose components may have the mass of 120 Suns — embedded within it. Elsewhere in Carina lie two other bright star clusters known as the Wishing Well (NGC 3532) and the Southern Pleiades (IC 2602).

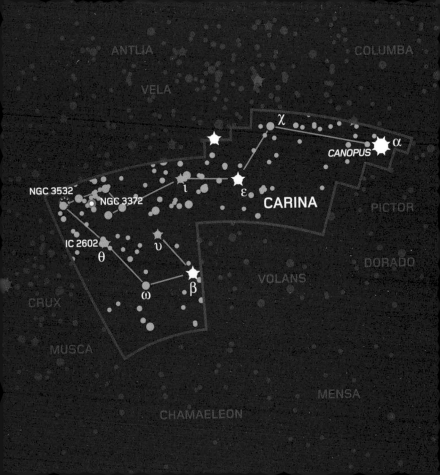

Crux and Musca

The famous Southern Cross is the smallest constellation in the sky, but also one of the brightest. The ancient Greeks knew its stars as part of Centaurus and it owes its modern identity to 16th-century European navigators. To its south lies Musca, invented around the same time as Crux, and once seen as a bee rather than a fly.

Acrux (Alpha Crucis), at the southern end of the cross, is a beautiful blue-white double star of magnitude 0.8, whose components, 320 light years away, are neatly separated by a small telescope. Theta Muscae is another double – apparently much fainter at magnitude 5.7, but in reality far more brilliant as we see its stars over a distance of about 10,000 light years. Gamma Crucis is a magnitude-1.6 red giant, 88 light years from Earth, which forms a stark contrast with its blue-white neighbours. Tucked into one corner of Crux are the compact dark nebula known as the Coalsack and a colourful naked-eye star cluster called the Jewel Box (NGC 4755).

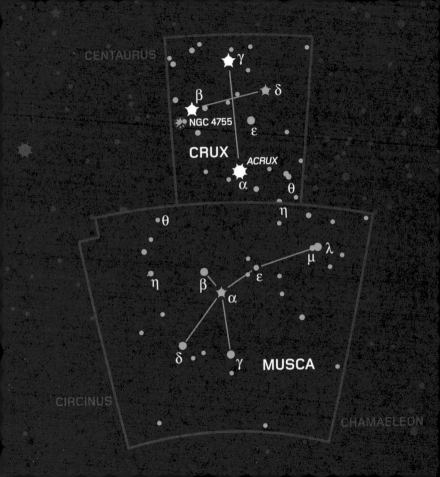

Circinus and
Triangulum Australe

The brighter stars of these two constellations both form triangles of different shapes. Circinus is long and thin, reminding its 18th-century inventor Lacaille of a set of Surveyor's Compasses. The broader equilateral shape of the Southern Triangle was apparently more noticeable to early navigators, since it was first recorded in the late 16th century.

Atria (Alpha Trianguli Australe) is an orange giant of magnitude 1.9, roughly 415 light years from Earth, while the nameless Alpha Circini is a binary some 53 light years away – its white primary (magnitude 3.2) has an orange-dwarf companion visible through a telescope. Lambda Circini, meanwhile, is actually not a single star but an open cluster, also designated as NGC 6025. This magnitude-5.1 'fuzzy star' dissolves into a cluster of 30 or more stars, 2,700 light years distant, through binoculars. Milky Way star clouds running through Circinus block the view of ESO 97-G13 (the Circinus Galaxy), an active galaxy just 13 million light years away, but only discovered in the 1970s.

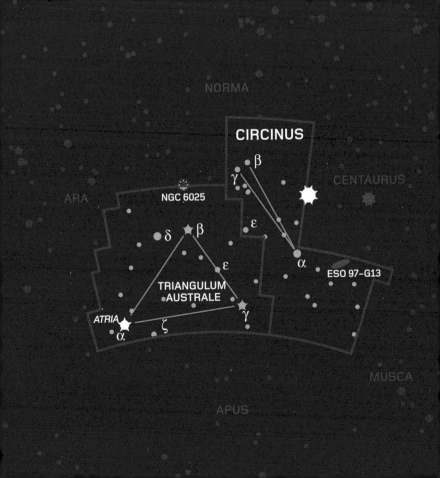

Norma and Ara

Despite its obscurity, the small constellation of Ara the Altar, to the south of Scorpius, is actually one of the original 48 classical constellations. Its neigbour Norma, in contrast, is another of the uninspiring space-fillers invented by Nicolas Louis de Lacaille in the 1750s.

Gamma Normae is a line-of-sight double consisting of yellow stars of magnitudes 4.0 and 5.0 at very different distances from Earth – the fainter star is a yellow supergiant 1,500 light years away, while the brighter one is a more normal yellow giant about 130 light years away. NGC 6087 is a small open cluster with an overall magnitude of about 5.6. Binoculars reveal a group of about 40 stars centred on another yellow supergiant – a Cepheid variable (see page 300) catalogued as S Normae. NGC 6193, meanwhile, is a slightly brighter cluster over the border in Ara, with two dozen or more members marking the centre of a broader 'OB1 association' of luminous blue and white stars originating in the nearby nebula NGC 6188.

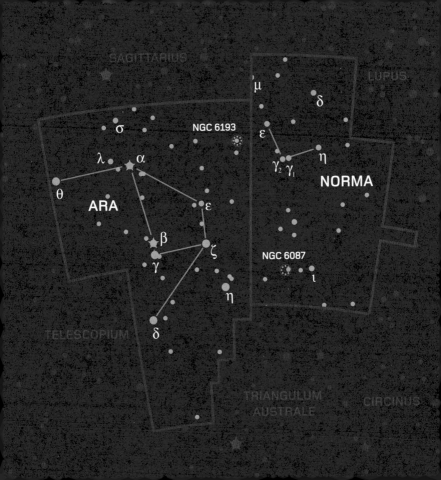

Corona Australis

The Southern Crown (counterpart to Corona Borealis – see page 150) is a faint arc of stars tucked away to the south of the brighter teapot-shaped body of Sagittarius, and east of the sting of Scorpius. Its shape is not as perfect as that of Corona Australis, but is nevertheless quite distinctive under dark skies. It was known to the ancient Greeks, who identified it with the wreath worn by Dionysus, the god of wine.

The constellation's brightest stars are Alpha and Beta Coronae Australis, both shining at magnitude 4.1. Alpha is a white star 140 light years from Earth, while Beta is a yellow giant some 510 light years away. Gamma is an attractive binary, easily separated into a roughly equal pair of yellow stars through a small telescope. The northern reaches of the constellation are dominated by the Corona Australis Molecular Cloud, one of the closest star-forming regions to Earth. Though just 430 light years away, it is unfortunately obscure since it is dominated by dark, dust-laden clouds rather than bright emission nebulae.

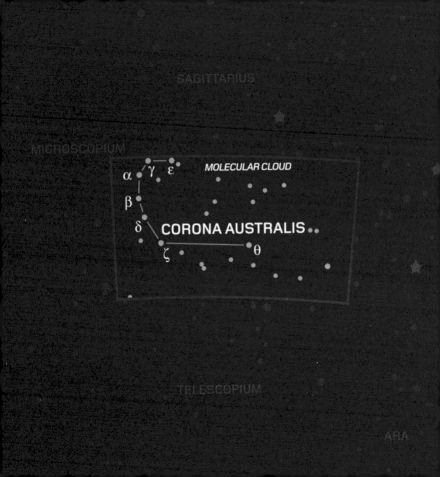

Pavo and Telescopium

These two faint constellations trace their origin to two different traditions in the division of southern skies. Pavo is one of several 'Southern Birds' based on the observations of Dutch navigators Pieter Dirkszoon Keyser and Frederick de Houtman, who made some of the first detailed observations of far southern skies during a trading expedition in the mid-1590s. Telescopium, meanwhile, is one of 14 late additions to the heavens made by Nicolas Louis de Lacaille in the 1750s, and may be the most obscure constellation in the entire sky.

Peacock (Alpha Pavonis) is a magnitude-1.9 blue-giant star 183 light years from Earth, which spectroscopic observations reveal is actually a tight double star. Kappa Pavonis is a bright example of a Cepheid variable (see page 300), 540 light years away and pulsating between magnitudes 3.9 and 4.8 in a 9.1-day cycle. NGC 6752, also in Pavo, is a beautiful globular cluster – the third brightest in the sky at magnitude 5.4 and a beautiful sight through any optical instrument.

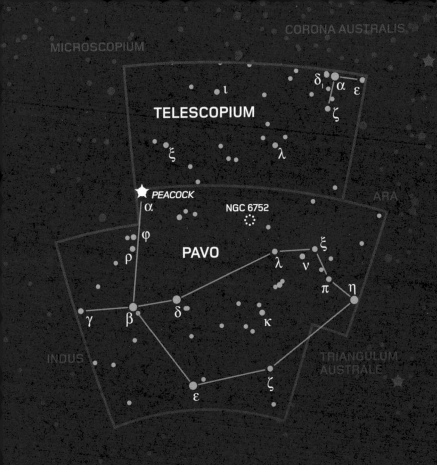

Grus and Phoenix

These two celestial birds are best found by looking to the south of the bright star Fomalhaut in Piscis Austrinus. Grus has at least a couple of stars of moderate brightness as well as a distinctive diagonal chain of stars representing its outstretched neck. Phoenix is fainter, with a broad shape that supposedly represents the outstretched wings of the mythical bird. Both were invented from the observations of Dutch navigators Keyser and de Houtman in the late 16th century.

Delta Gruis is a naked-eye double star that is merely a line-of-sight effect caused by yellow and red giants, of magnitudes 4.0 and 4.1, lying in the same direction in the sky, but at wildly different distances of 150 and 420 light years. Zeta Phoenicis, however, is a genuine multiple star – a quadruple whose brightest member, an eclipsing binary, dips from magnitude 3.9 to 4.4 every 40 hours. Small telescopes reveal a third component of magnitude 6.9, but larger ones are needed to see the fourth member, fainter and closer to the primary.

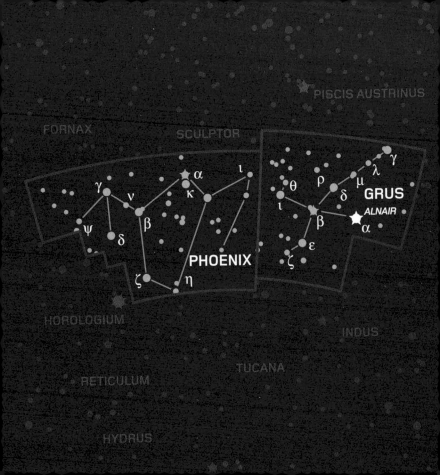

Dorado, Pictor and Reticulum

The constellation of the Dorado (a fish known as the *mahi-mahi* in the East Indies) was added to the sky by Dutch navigators Keyser and de Houtman in the late 16th century, perhaps based on native folklore. It is flanked by two much later neighbours – the Easel and the Reticle (a specialized type of eyepiece) invented by Nicolas Louis de Lacaille in the 1750s.

Magnitude-3.9 Beta Pictoris is an unremarkable white star 63 light years from Earth, around which infrared telescopes have discovered a disc of planet-forming material. However, the area's main draw for observers is the Large Magellanic Cloud (see page 350), a satellite galaxy in orbit around our own. From a distance of 160,000 light years, it looks at first glance like an isolated section of the Milky Way, running across southern Dorado into neighbouring Mensa. Binoculars or a telescope will reveal individual features within it, the most impressive of which is the Tarantula Nebula (NGC 2070), the largest star-forming nebula in our cosmic neighbourhood.

Mensa and Volans

Although its official name means simply 'the Table', the small far-southern constellation of Mensa is actually unique as the only star group named after a geographical feature. Invented by French astronomer Nicolas Louis de Lacaille in the 1750s, it was originally Mons Mensae, honouring a supposed resemblance between its quadrilateral of stars and South Africa's Table Mountain. The Large Magellanic Cloud, spilling over from Dorado, forms a misty patch over the 'mountain's' flat top. Nearby Volans, meanwhile, is one of several groups added to the sky by Dutch navigators in the late 16th century.

Alpha Mensae is an unremarkable star of magnitude 5.1 with little to recommend it aside from its similarity to our own Sun: it lies 33 light years from Earth and is estimated to have 90 per cent of the Sun's mass and shine with 80 per cent of its luminosity. Gamma Volantis, meanwhile, is a beautiful binary for small telescopes, consisting of a magnitude-3.8 orange giant with a white companion of magnitude 5.7, 150 light years away.

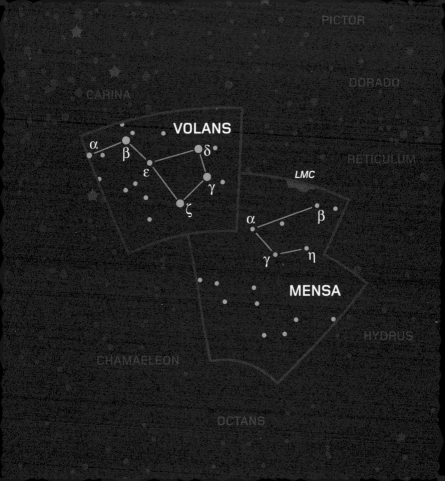

Chamaeleon and Apus

These two faint constellations, sandwiched between the bright stars of Carina, Crux and Centaurus, and the barren region of the south celestial pole, are both inventions of the Dutch navigators Pieter Dirkszoon Keyser and Frederick de Houtman, although they were first recorded by astronomer Petrus Plancius in the late 16th century.

Delta Chamaeleontis is a line-of-sight double star whose components are easily separated in binoculars. The brighter star, shining at magnitude 4.4, is a blue-white 'subgiant' 780 light years from Earth, while the fainter element is an unresolvable binary pair of orange giants about 350 light years away, with a combined brightness of magnitude 5.5. Coincidentally, Delta Apodis is also a double star, though in this case the two stars (orange giants of magnitudes 4.7 and 5.3) are a genuine gravitationally bound pairing. Despite a distance of 310 light years, their separation is so wide that they can be easily split with binoculars, or even the unaided eye.

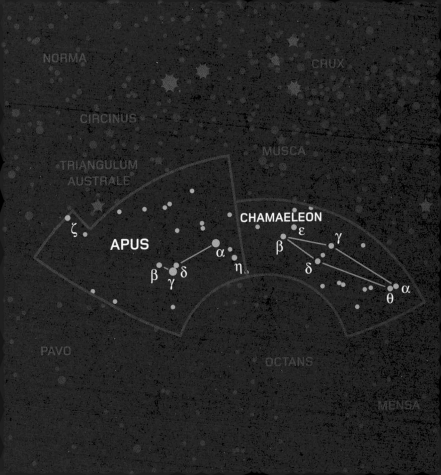

Indus and Tucana

Although often represented as a Native American complete with feathered headdress, the faint constellation of Indus more likely represents a native of the southeast Asian islands visited by Dutch navigators Keyser and de Houtman in the 1590s. This pair were also responsible for naming the nearby and somewhat brighter Toucan.

Indus has little to see except for magnitude-4.7 Epsilon Indi, a Sun-like star just 11.2 light years from Earth and known to have a 'brown-dwarf' companion 45 times the mass of Jupiter. Tucana, in contrast, is home to the beautiful Small Magellanic Cloud – a small irregular galaxy (see page 358) orbiting the Milky Way at a distance of about 210,000 light years. Binoculars or small telescopes will reveal many interesting features within it. Nearby, the globular star cluster 47 Tucanae is much closer to Earth – about 17,000 light years away. Just 120 light years across, but packed with perhaps half a million stars, it appears to the naked eye as a fuzzy 'star' of magnitude 4.9.

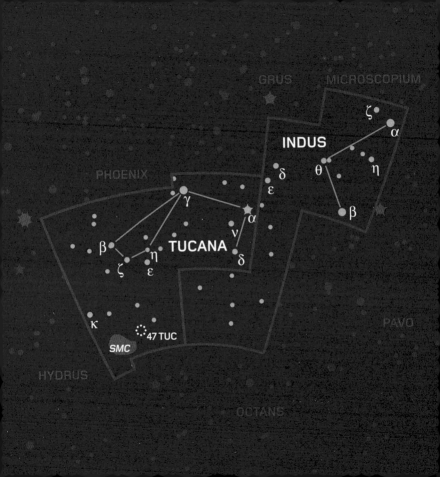

Hydrus and Octans

In contrast to the bright constellations of the north celestial pole, the region around the south celestial pole is barren of obvious landmarks, let alone a bright 'pole star'. The Little Water Snake (not to be confused with the huge Hydra of equatorial skies) winds its way south towards the pole from beneath Achernar in Eridanus, and was recognized by Dutch navigators in the late 16th century. Nicolas Louis de Lacaille was responsible for the invention of the obscure Octant, which marks the celestial pole itself.

Alpha and Beta Hydri, on opposite corners of the constellation, are of almost equal brightness at magnitudes 2.9 and 2.8 respectively. Alpha is a white star 70 light years from Earth, while Beta, the nearest moderately bright star to the pole, is a yellow 'subgiant' – just 24 light years away and beginning its evolution towards becoming a red giant (see page 312). The closest naked-eye star to the pole (roughly 1 degree away) is magnitude-5.4 Sigma Octantis.

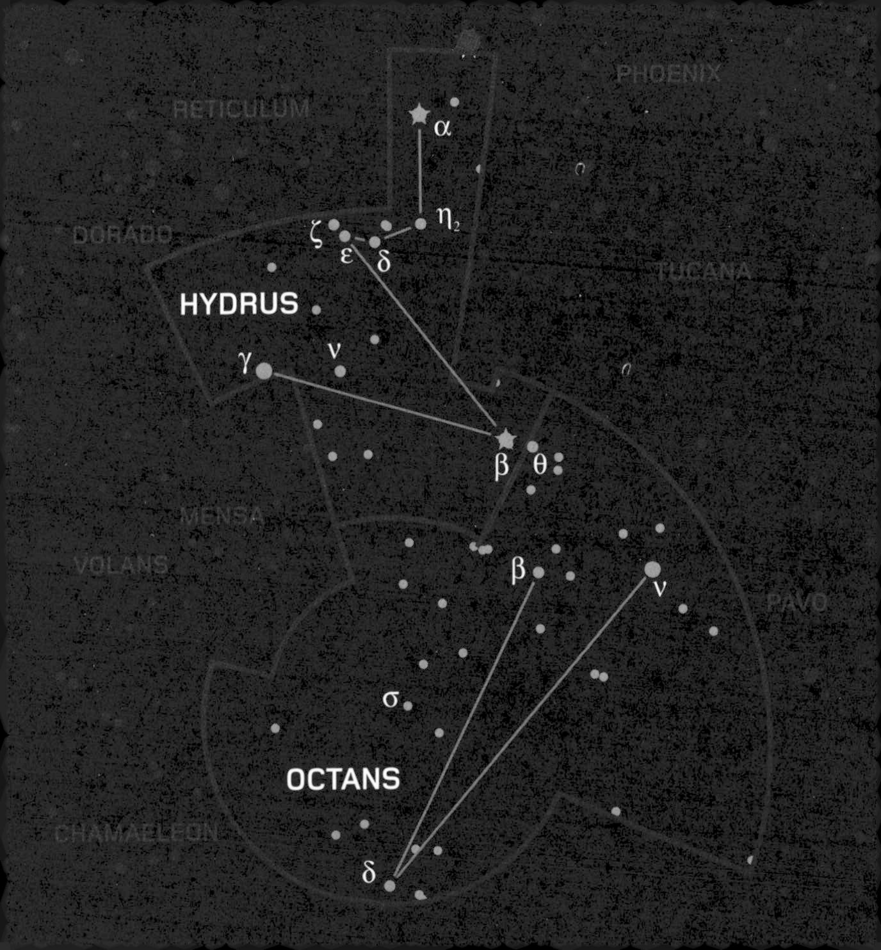

RETICULUM
PHOENIX
α
η2
ζ
ε
δ
DORADO
TUCANA
HYDRUS
γ
ν
β
θ
MENSA
VOLANS
β
ν
PAVO
σ
OCTANS
CHAMAELEON
δ

The lives of stars

Stars are the most important objects in the Universe – huge balls of gas that grow dense and hot enough to sustain nuclear fusion in their core. They are the ultimate source of all high-energy radiation such as visible light, and also the means by which the lightweight raw materials of the Universe (see page 394) are transformed into heavier and more complex elements – including those that make up our own bodies.

Under clear, dark skies, a person with good eyesight can typically see about 3,000 stars at any one time. Two key differences among them are instantly apparent – variations in brightness ('apparent magnitude') and colour ('spectral type'). Such differences reflect genuine variety in the intrinsic properties of the stars themselves, but the distances of the naked-eye stars vary hugely, from a few light years to many thousands, and we should not make the mistake of assuming that the brightest stars in our sky are the most luminous in reality.

In fact, the sample of naked-eye stars in Earth's skies is inevitably skewed by such factors, and is certainly not representative of our galaxy's stellar population as a whole. Highly luminous giant stars are actually very rare, but their brightness makes them prominent over vast distances, while fainter dwarfs are much more numerous, but only a handful of those closest to Earth are visible to the naked-eye at all. Attempts to understand the true distribution of stellar properties in the early 20th century led to the development of the Hertzsprung-Russell diagram (see page 268), which is key to our modern understanding of the evolution of stars.

The stellar life cycle is one of an unstable youth, an extended middle age, a bloated old age and a death that may be either gentle or spectacular. The entire cycle may proceed at wildly different speeds depending on the composition and mass of the star involved. Stars like our Sun are stable for 10 billion years, while those a few times more massive may burn through all their fuel in just a few million years – perhaps $1/1,000$ of the Sun's lifespan. At the other extreme, dwarf stars with perhaps half the Sun's mass may shine feebly, but will remain essentially unaltered for far longer than the current age of the Universe itself.

Stellar distances

Perhaps the most basic challenge to understanding the properties of stars and other celestial objects is the accurate measurement of their distance. Combined with the apparent magnitude (its brightness as seen from Earth), knowledge of a star's distance can reveal its inherent energy output, or luminosity, the key to its other physical properties.

Since we cannot travel to other stars in order to determine their distance, astronomers rely on a number of other techniques. The most accurate, known as the parallax method, involves measuring the slight shift in the direction of nearby stars as Earth moves from one side of its orbit to the other. Modern satellites can measure the tiny parallax shifts of objects hundreds or thousands of light years away from Earth. Once true luminosities are determined, other relationships become apparent, revealing 'rules of thumb' by which astronomers can estimate the distance and luminosity of objects far beyond the direct use of parallax.

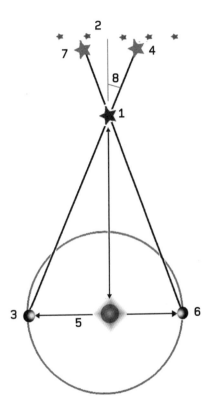

Using parallax to measure the distance to stars

1 Foreground star close to the solar system

2 Distant background stars

3 Earth at time of first measurement

4 Apparent position of star in first measurement

5 Opposite sides of Earth's orbit, 300 million km (186 million miles) apart

6 Earth at second measurement, six months later

7 Apparent position of star in second measurement

8 'Parallax angle' reveals distance from Earth to foreground star

Star colour and brightness

Stars in the night sky vary in two obvious ways – brightness and colour – and the ability to measure stellar distances confirms that the true brightnesses, or luminosities, of stars vary just as much as their apparent magnitudes in the sky. Some stars are tens of thousands of times less luminous than our Sun, while others are millions of times brighter.

Colour, meanwhile, is an indication of a star's surface temperature: yellow stars like the Sun have surface temperatures around 5,800°C (9,900°F), while orange and red stars are thousands of degrees cooler, and white and blue stars far hotter. Because a star's temperature is an indication of the energy passing through each unit of its surface, it follows that for cool red and hot blue stars of identical luminosities, the cooler red star must be considerably smaller so that the energy escapes through a reduced surface area and has a far greater heating effect. Hence, knowledge of a star's colour and luminosity can reveal its size.

THE SUN a yellow dwarf

RED AND ORANGE DWARFS

BETELGEUSE a red supergiant

POLLUX a yellow giant

SIRIUS a white main-sequence star

RIGEL a blue giant

ALDEBARAN an orange giant

Stellar chemistry

A star's colour arises from the precise distribution of its
energy output across different wavelengths, which in
turn is affected by its surface temperature. However, closer
analysis of light from stars and other celestial objects, using
the techniques of spectroscopy (see page 24), can reveal the
intimate secrets of their chemical make-up.

Astronomical spectra typically come in two forms. Stars
reveal absorption spectra, in which a rainbow-like 'continuum'
of light is crossed by narrow dark lines indicating that light
is being absorbed at specific wavelengths. Other phenomena
show emission spectra that are mostly dark with a few specific
bright lines emitted at particular energies. Both types of
spectrum can reveal the presence of specific atoms and
molecules, which change their internal structure in order
to absorb or emit energy. They act as chemical 'fingerprints'
for studying the composition of objects and also reveal the
presence of intervening material that affects their light.

Spectroscope

Incandescent star surface produces a continuous spectrum

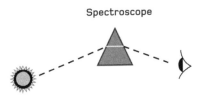

Star

Glowing interstellar gas generates emission spectrum with lines of specific wavelengths

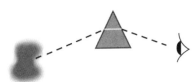

Hot gas

Cool gas in a star's outer layers absorb specific energies from its light to create an absorption spectrum of dark lines.

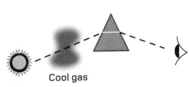

Star   Cool gas

Stars in motion

Even with powerful modern telescopes, it is difficult to detect the movement in space of any but the most rapidly moving stars. And even those stars that do move significantly reveal only their lateral (sideways) motion across the sky. However, the tightly defined absorption lines embedded in stellar spectra (see page 282) reveal movement in another way.

When a star has 'radial' motion towards or away from Earth, its light experiences a Doppler shift – the same effect that changes the pitch of sound waves when an emergency siren passes by in the street. Light waves from objects moving towards Earth are compressed, while those from retreating sources are stretched, producing effects known as blue shift and red shift. Doppler shifts affect all of a star's light equally, but if astronomers can identify specific patterns of spectral lines that have moved slightly from their expected position, this can reveal a star's radial motion, revealing, for example, its orbital period and speed within a binary system (see page 298).

Star moving away
from Earth

Wavelengths of light stretched

Earth

Spectral lines shifted towards red

Star moving
towards Earth

Wavelengths of light shortened

Earth

Spectral lines shifted towards blue

Measuring stellar masses

One of the most important stellar properties is mass. The quantity of material within a star dictates the amount of hydrogen available to burn and (by controlling the pressure and temperature in the core) the rate at which it burns through that fuel and hence the star's lifespan and luminosity.

The most direct way of measuring a star's mass is through the simple laws of orbital mechanics. The orbital periods of stars in a binary or multiple star system (see page 298), reveal their proximity to the system's centre of mass, and therefore their comparative masses – heavier stars will follow smaller orbits closer to the centre of mass than lighter ones. Knowledge of the precise dimensions and orientation of a binary system would allow a calculation of the absolute mass of the stars involved but, in practice, measurements are limited and astronomers can usually estimate only relative masses. Nevertheless, this is still enough to reveal important relationships between stellar mass and luminosity.

The size of orbits in a binary system reveal the relative masses of its stars, allowing comparisons of the way that mass affects other stellar properties.

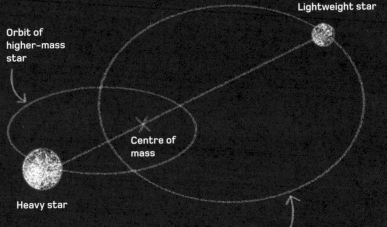

Lightweight star

Orbit of higher-mass star

Centre of mass

Heavy star

Orbit of lower-mass star

The Hertzsprung–Russell Diagram

Developed independently by Sweden's Ejnar Hertzsprung and American Henry Norris Russell around 1911, the Hertzsprung-Russell (H-R) diagram plays a key role in understanding stellar evolution. Plotting the positions of stars in terms of their colour or surface temperature and their luminosity, it reveals a clear pattern in their distribution. The vast majority of stars lie along a diagonal band known as the 'main sequence', which links cool, faint stars (red dwarfs) with hot, bright ones (blue giants). A relatively small number of stars occupy other parts of the diagram – faint but hot white dwarfs, cool but brilliant red giants, and a band of incredibly luminous supergiant stars of all colours. The overwhelming dominance of the main sequence shows that stars spend most of their lives here. Giants and supergiants are unusually prominent in the sky because their great luminosity makes them visible across huge distances, but based on surveys of our closest stellar neighbours, dwarf stars are actually the most common by far.

Stellar groups on the Hertzsprung–Russell diagram

1 Main sequence of stars burning hydrogen in their cores – position along line depends on mass of star.

2 Red and orange giants are expanding bright stars near the end of their lives.

3 Most massive stars swell into supergiants as they age.

4 White dwarfs are hot but faint cores left behind when Sun-like stars burn out.

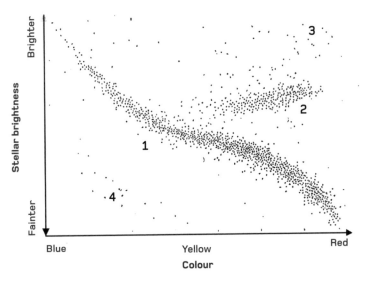

Interpreting the H–R diagram

One early interpretation of the famous Hertzsprung–Russell diagram (see page 268) involved stars moving along the main sequence over the course of their lifetimes, either becoming hotter and brighter, or cooler and fainter. In 1924, however, British astronomer Arthur Stanley Eddington showed (by studying stars of different masses in binary systems) that for the great majority of stars, mass and luminosity are closely linked. Stars with tens of solar masses may be millions of times brighter and hotter, while those with a fraction of the Sun's mass are thousands of times fainter and cooler.

Eddington was also the first person to suggest that stars shine by nuclear fusion (see pages 272–5) – a process that means they maintain roughly the same mass throughout their lifetimes. Most stars therefore occupy more or less the same position on the main sequence for the majority of their lives, only moving away from it as they expand and brighten into red giants (see page 312) once the fuel in their core is exhausted.

Evolution of a Sun-like star

1 Through most of its life, the star's position on the main sequence is determined by its mass.

2 As the star exhausts its core hydrogen, it swells, brightens and cools, becoming a red giant.

3 The star makes a small 'loop' while burning helium in its core, then swells once again when this is used up.

4 Finally the star sheds its outer layers to expose the faint but hot burnt-out core – a white dwarf.

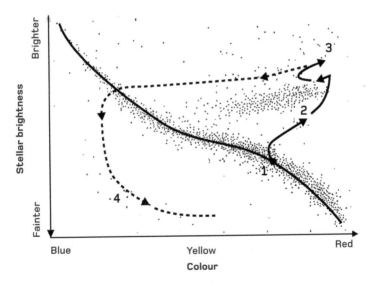

Nuclear fusion: the PP chain

Stars spend the majority of their lives generating energy by the nuclear fusion of hydrogen (the lightest and most abundant element in the Universe) into helium (the next lightest and second most abundant) in their cores. Depending on the temperature, pressure and availability of other elements, this process can happen via either the proton-proton ('PP') chain or the carbon-nitrogen-oxygen ('CNO') cycle (see page 274).

For low-mass stars like the Sun, the PP chain is dominant. This involves the collision and fusion of protons (stripped-down hydrogen nuclei) to build heavier atomic nuclei. In the process, protons spontaneously transform into uncharged neutrons, emitting tiny, near-massless particles known as neutrinos. The resulting helium nucleus has slightly less mass than the four hydrogen nuclei used to create it, and the excess is released as energy in the form of gamma rays. The rays heat the star's interior and slowly make their way to the surface, transforming into lower-energy radiation along the way.

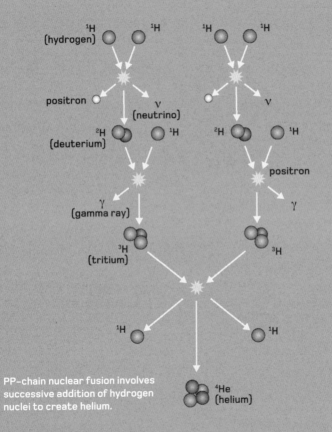

PP-chain nuclear fusion involves successive addition of hydrogen nuclei to create helium.

Nuclear fusion: the CNO cycle

In stars more massive than the Sun, high temperatures and pressures in the core allow a second nuclear fusion process to dominate over the simple low-temperature 'proton-proton chain'. The CNO cycle relies on the presence of small numbers of carbon nuclei in the stellar core. This means it can only take place in stars that contain significant amounts of material recycled from previous stellar generations (see page 326). Protons fusing with the carbon create nuclei of the heavier elements nitrogen and oxygen, before addition of a further proton causes the nucleus to break apart, releasing helium and unaltered carbon. The CNO cycle is much more efficient than the random collisions of the PP chain, fusing hydrogen to make helium much more quickly. It becomes ever more efficient at higher temperatures and pressures, allowing the most massive stars to pump out millions of times more energy than their fainter cousins. The downside of this is that they squander their core supplies of fuel over very short timescales, with dramatic consequences for their later evolution.

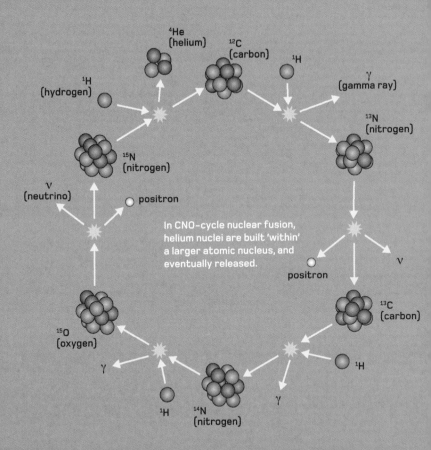

In CNO-cycle nuclear fusion, helium nuclei are built 'within' a larger atomic nucleus, and eventually released.

The interstellar medium

The space between the stars is not empty – within our galaxy and others, it is filled with the scattered gas and dust of the interstellar medium. In places, this concentrates to form denser clouds, known as nebulae, that reveal themselves in three ways. Dark nebulae are dusty, light-absorbing clouds that become visible in silhouette against the brighter background of another nebula or cloud of stars. Reflection nebulae shine by reflecting or scattering the light from nearby stars: they often have a bluish colour, since the process that illuminates them is rather similar to that which scatters sunlight to make Earth's daylit skies blue. Finally, emission nebulae are fluorescent clouds, glowing at specific wavelengths as their atoms and molecules become excited by high-energy ultraviolet rays from nearby stars, and then return to their unexcited state by emitting visible light. Emission nebulae can display a variety of colours depending on their chemical composition, but the vast majority, dominated by hydrogen and associated with star formation, have a distinctive pinkish hue.

The famous Horsehead Nebula in Orion is a dark dust column silhouetted against a more distant emission nebula.

Star formation

Stars are born when an external influence triggers the collapse of an already-dense cloud of interstellar medium. The process can be started by the shock wave from a nearby supernova (see page 324), tides raised by a close encounter with an older star or passage through a 'spiral density wave' (see page 344). Often, the result is a dense, opaque column or pillar of starbirth, perhaps a dozen light years long and containing several distinct knots of collapsing material – individual 'protostars' whose gravity eventually grows powerful enough to pull in more material from their surroundings. Such clouds usually display a small amount of random rotation, and as the protostar grows more massive and starts to pull material closer to its centre, it spins more rapidly thanks to the conservation of angular momentum, flattening out into a spinning disc with a bulge at its centre. The core of the protostar continues to get denser and hotter, until eventually it becomes hot enough to trigger nuclear fusion in its core – at which point, it has become a star.

The 'Pillars of Creation' in the Eagle Nebula (M16 in Serpens) are opaque columns of collapsing gas and dust within which stars are being born.

Young stars

Stars take some time to settle down onto the 'main sequence' of stellar evolution. As they contract over millions of years, they radiate at first through heat generated by gravitational contraction, with the increasing temperature and density in their core causing them to brighten over time even before nuclear fusion begins. During this phase, they may be prone to sudden outbursts and eruptions – low-mass stars of this type are called T Tauri stars, while higher-mass stellar infants are called Herbig Ae and Be stars.

As the star grows smaller and denser, it continues to pull in more material from its surroundings. This forms a flattened 'accretion disc' spiralling down onto the star, with fast-moving gas ejected from above and below the disc in a phenomenon called 'bipolar outflow'. The gas forms jets that may extend for a light year or more, before billowing out into bullet-shaped shock waves as it encounters nearby interstellar gas – a phenomenon known as a 'Herbig-Haro object'.

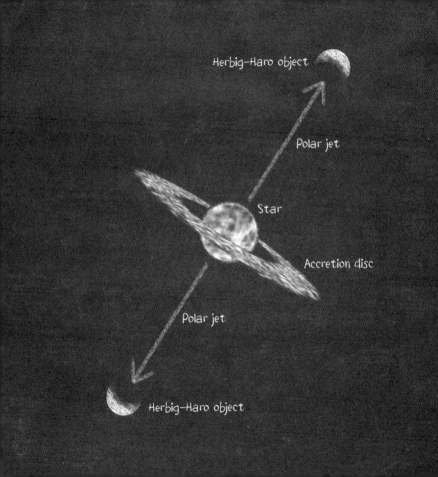

Brown dwarfs

In recent decades, astronomers using powerful Earth- and space-based telescopes have discovered a new type of star that had previously escaped their attention. Brown dwarfs are objects that form in the same way as true stars, but lack the mass required to trigger hydrogen fusion in their cores. Instead, they emit energy largely in the infrared, through a mixture of gravitational contraction and the fusion of deuterium (a rare 'isotope' or form of hydrogen that fuses more easily than normal hydrogen).

Brown dwarfs have an upper mass limit of 0.075 Suns or 75 Jupiters, above which they become red dwarfs (see page 284). Their lower mass limit is less well defined, and the boundary between giant planets and small brown dwarfs is blurred. Remarkably, whatever their mass, brown dwarfs always have about the same diameter as Jupiter. Infrared surveys suggest that they are abundant in the nearby galaxy, and some even have planets of their own in orbit around them.

This schematic depicts our Sun, a typical brown dwarf and a Jupiter-like gas giant to scale.

The Sun

200,000 km
(125,000 miles)

Brown dwarf

Gas giant planet

Red dwarfs

The least massive of fully fledged stars, red dwarfs are also the most common, accounting for perhaps three-quarters of all stars in the Milky Way. They range from around 0.075 to 0.5 solar masses, above which stars become much more Sun-like in their behaviour. Their luminosities, meanwhile, range from about $1/15$ of the Sun's down to around $1/10,000$ making them extremely faint and hard to detect even in our solar neighbourhood.

The internal structure of red dwarfs is quite simple, with energy transported from the core to the surface by large-scale convection (hot material expanding and pushing its way up through cooler overlying layers). This process constantly refreshes the core with fresh hydrogen from the overlying layers and prevents the build-up of helium that ultimately causes burn-out in heavier stars like our Sun. As a result, red dwarfs have astonishingly long lifetimes – they should shine more or less unchanged for hundreds of billions of years.

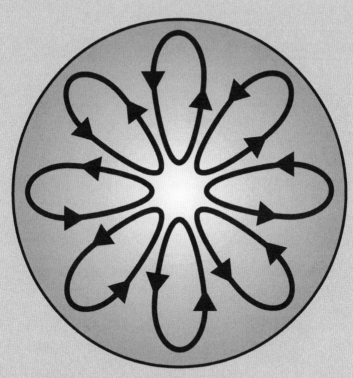

In a low-mass star, such as a red dwarf, convection occurs throughout the stellar interior.

Sun-like stars

Sun-like stars are perhaps the best understood, thanks to our ability to study the Sun in detail. While on the main sequence, they have an internal structure with three distinct layers: a core in which the energy is generated; a 'radiative zone' in which energy is transferred towards the surface as electromagnetic radiation, losing energy as it goes; and an upper convective zone in which energy is moved to the surface through bulk heating and movement of large masses of material. At the star's photosphere, its material becomes transparent, allowing energy to escape into space as electromagnetic radiation (mostly infrared and visible light).

These stars have long but finite lives – the Sun is halfway through an estimated 10-billion-year main-sequence lifespan during which it burns hydrogen, mainly through PP-chain nuclear fusion. Once core supplies of hydrogen are exhausted, changes to the star's internal structure cause it to brighten and swell, transforming it into a red giant (see page 312).

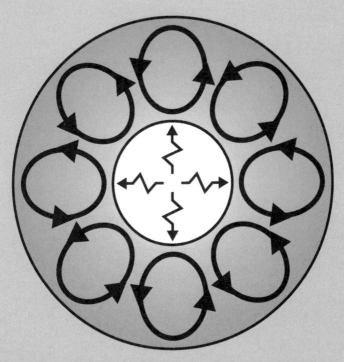

In an intermediate-mass star like the Sun, radiation carries energy away from the core before convection takes over in the upper layers.

Main-sequence heavyweights

The heaviest stars are comparatively rare in the Universe, but are well represented among the naked-eye stars because their brilliance allows them to be seen over huge distances. The most massive stars of all have such high surface temperatures that, while they appear blue-white in colour, they actually emit a great deal of their radiation as invisible ultraviolet rays.

Blue giants shine predominantly through the CNO cycle (see page 274), which allows them to squander enormous amounts of fuel in just a few million years and explains their high luminosities. The cycle's sensitivity to temperature is responsible for an internal structure that is rather different from those inside Sun-like stars. It produces convection cells deep inside the star, around the core itself. The convection zone is overlaid with a hot, 'foggy' radiation zone in which photons of energy bounce between densely packed atomic nuclei as they force their way out towards the surface.

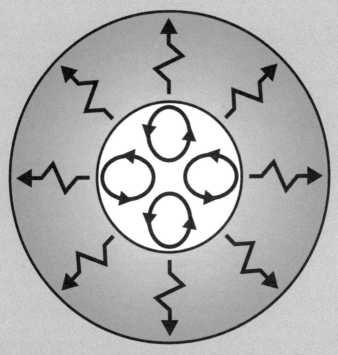

In a high-mass blue star, convection carries energy away from the core, but above a certain level, radiative transfer takes over.

Extrasolar planets

Since the 1990s, astronomers have finally begun to discover planets in orbit around other stars – and as telescopes have grown more powerful and techniques more sophisticated, the rate of discovery has snowballed.

Because planets are so small compared to their stars, and shine very feebly in reflected starlight, they are very hard to observe directly through even the largest telescopes – instead, a variety of other techniques are used. The first to prove its worth, and still the most successful, was the 'radial velocity technique', which uses tiny Doppler shifts in starlight to detect the slight wobbles in a star's position caused by orbiting planets (see page 264). Another highly successful technique is the 'transit method', which detects tiny dips in starlight caused when a planet moves across the face of its star. All techniques have their advantages and limitations, but the result is that more than a thousand extrasolar planets are now known, and the number continues to grow rapidly.

The radial velocity technique

1 Star

2 Planet

3 Star and planet both orbit shared centre of gravity, so the star 'wobbles' in space.

4 Change in apparent position of the star is too small to detect from Earth, but back-and-forth wobble can be measured from the Doppler shift of its light.

The variety of planets

The discovery of the first extrasolar planets has led to the realization that our own apparently orderly solar system is an exception rather than the norm. Several entirely new classes of planet have now been identified, the most important of which are probably the 'hot Jupiters' – giant planets orbiting their stars in a matter of days or even hours. Such worlds cannot possibly have formed so close to a star, so it seems they must have spiralled inwards over time, probably disrupting the orbits of inner worlds as they passed. Several hot Jupiters appear to have lost large amounts of their atmosphere to evaporation, and there are even signs of planets that have collided with their stars in the past.

Other exceptions to the familiar rules of our solar system include the existence of planets in highly elliptical orbits, and the existence of apparently solid 'super-Earths' many times larger than the rocky planets of our solar system. Planets have even been found in stable orbits around binary star systems.

Life in the Universe

The existence of extrasolar planets raises the obvious question of whether such worlds could support life. The science of 'astrobiology' is in its infancy, but the search for potentially habitable planets is geared to identifying Earth-like planets orbiting in the 'Goldilocks zone' around long-lived stars – the region where water can potentially exist on their surface as a stable liquid.

There are solid reasons to believe that life elsewhere in the Universe would rely on carbon-based 'organic' chemistry and require the presence of liquid water: carbon is uniquely suited to forming the large, complex molecules essential to life, while water is the best and most probable solvent for allowing molecules to move around and undergo chemical reactions. However, the discovery of 'extremophile' organisms in apparently hostile environments on Earth, coupled with the evidence for liquid water among the moons of Jupiter and Saturn, shows the potential for life in unexpected places.

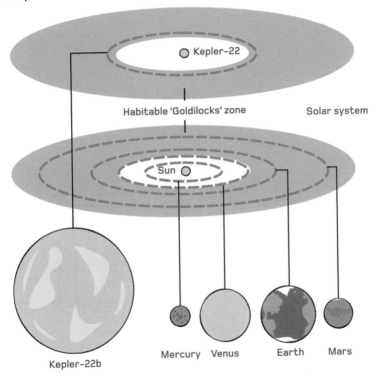

The first planet to be discovered in an orbit suitable for life is Kepler-22b, shown here in comparison with our own inner solar system.

Kepler-22 system

Kepler-22

Habitable 'Goldilocks' zone

Solar system

Sun

Kepler-22b

Mercury   Venus   Earth   Mars

Extraterrestrial intelligence

In statistical terms, the existence of intelligent life elsewhere in the Universe is more or less a certainty. But estimating its abundance, and attempting to locate and even communicate with it, present huge challenges for astronomers involved in the Search for Extraterrestrial Intelligence (SETI) programme.

Rate of star formation in our galaxy	Proportion of stars with planetary systems	Number of habitable planets per system	Proportion of planets with life

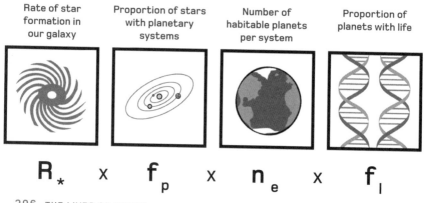

$$R_* \quad \times \quad f_p \quad \times \quad n_e \quad \times \quad f_l$$

The Drake Equation (shown below) offers a potential means of assessing the numbers of 'communicating civilizations' in our galaxy, but with only one example of such a civlization to work from (our own), our understanding of the factors involved is currently very limited. More practical SETI efforts involve scanning the sky for weak radio signals sent from remote civilizations, or looking for telltale 'signatures' of advanced life, such as artificial pollutants in the atmospheres of extrasolar planets or planetary-scale artificial structures. To date, just one attempt has been made to deliberately send a radio signal with information about Earth into the wider Universe.

Proportion of life evolving intelligence	Proportion of intelligent life developing technology	Average lifetime of a communicating species	
			= **Number of detectable civilizations**
x **f_i**	x **f_c**	x **L**	

Multiple stars

A significant majority of stars in the Milky Way are members of binary or multiple star systems – groups that arise naturally from the processes of star formation. Stars can take anything from hours to millions of years to orbit each other, and may vary hugely in mass, size and brightness. Sometimes their individual components are easily separated through a telescope, but distant stars in tight orbits only reveal their true nature through studies of their spectra, and specifically the telltale Doppler shifts caused by their orbital motion (see page 264). Such systems are known as spectroscopic binaries.

Binaries and multiples can reveal important stellar properties. The size of each star's orbit around their shared 'centre of mass' is a direct indication of their relative masses, with the less massive members following larger orbits. Furthermore, multiple systems allow direct comparison of properties such as stellar brightness, since factors such as distance and light absorption by interstellar dust affect all their stars equally.

Computer simulation of
multiple star formation

Pulsating variables

A surprisingly large number of stars vary in their apparent magnitude as seen from Earth. Sometimes this is because they are actually binary systems (see pages 298 and 302), but the majority of variables really are changing their output of energy, in smoothly varying cycles that may show clockwork regularity or infuriating unpredictability. In most cases this change in brightness is matched by a change in physical size.

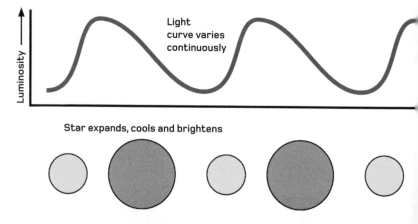

Luminosity →

Light curve varies continuously

Star expands, cools and brightens

Perhaps the most famous examples are the red-giant long-period variables such as Mira (see page 194) and the yellow supergiant Cepheids (see page 140).

Pulsations can develop at various stages in a star's life cycle, but most are linked to the same basic feedback mechanism, in which rising temperatures within the star cause it to brighten, and the force of escaping radiation forces it to expand. This allows the interior to cool, reducing the star's brightness and allowing it to contract. This in turn raises the internal temperature and causes the cycle to repeat.

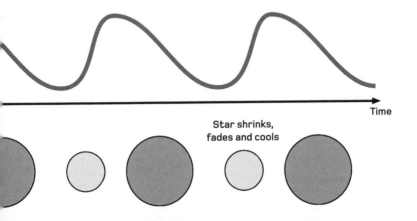

Time

Star shrinks,
fades and cools

Eclipsing binaries

In rare circumstances, binary stars have close relationships that cause their overall light output to vary periodically when seen from Earth. The most common situation is known as an 'eclipsing variable', and involves a coincidental alignment of the two stars orbits, so that, seen from Earth, they pass in front of and behind each other once in each orbit.

In order to produce an eclipse, the stars must be close enough to appear as a single object through even the most powerful telescope. The combined light of the two stars gives the system a steady brightness throughout much of each eclipse, with a sudden drop in brightness during the periods when one star passes in front of the other and blocks much of its light. Because the eclipsing stars are rarely of equal brightness, there is usually one much deeper 'primary eclipse' in which the fainter star moves in front of the brighter one, and a relatively shallow (sometimes unnoticeable) secondary eclipse in which the situation is reversed.

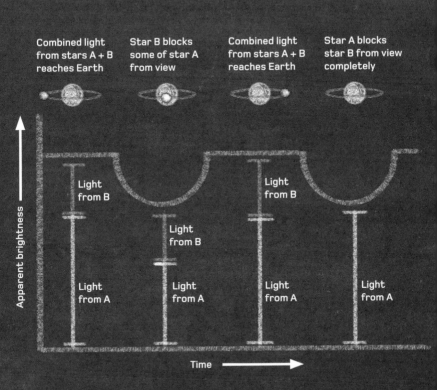

Asymmetric and eruptive variables

In addition to the pulsating and eclipsing types (see pages 300 and 302), there are several other types of variable star. Rotating variables change their brightness for a variety of reasons as they spin on their axes. Some stars appear to have distinctly asymmetric markings. These dark 'starspots' are much larger than the sunspots of our own Sun, and big enough to alter the star's overall light output when facing Earth. Stars in close binary systems, meanwhile, may distort each other's shapes so that they present different surface areas to Earth as they orbit. Intense magnetic fields can also cause a star to emit varying amounts of radiation in different directions.

Eruptive variables are unstable stars that are prone to sudden outbursts: they range from low-mass red-dwarf 'flare stars' that brighten when they emit huge stellar flares, to giant 'luminous blue variables', such as Eta Carinae (see page 322). The most violent eruptions of all – novae and supernovae (see pages 336, 324 and 338) – are known as cataclysmic variables.

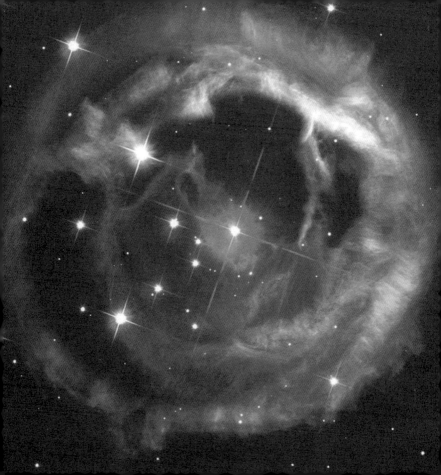

Open star clusters

Because stars form in large numbers over relatively short periods of time, they naturally emerge from their starbirth nebulae in groups known as open clusters. These relatively loose star clusters, containing anything from a few dozen to a few hundred stars, usually drift apart over tens of millions of years (though sometimes we can trace the middle-aged stars of a 'moving group' back to a common origin – see page 146). Clusters allow a useful direct comparison of the evolutionary paths taken by different stars, since all their stars are of the same age and distance, and formed from similar materials.

Since the majority of open clusters are so young, they tend to be dominated by luminous blue and white stars – heavyweights with lifespans of just a few million years that will exhaust their fuel and burn out before they can drift away from their neighbours. Such concentrations of luminous stars make open clusters easy to spot over huge distances – for instance, in the spiral arms of galaxies (see page 344).

Globular clusters

Vast balls containing hundreds of thousands of stars in a space perhaps 100 light years across, globular clusters are very different from the more common open clusters (see page 306). They inhabit different regions of our galaxy, either close to the galactic hub, or in the halo region above and below the flattened disc of our galaxy. And while open clusters comprise stars that have formed in the recent past, and are dominated by short-lived blue and white stars, globulars appear to be very ancient. Any heavyweight stars they once had have long since aged and died, leaving only low-mass red and yellow stars. (Rare 'blue stragglers' found in the cluster cores are thought to form when tightly packed stars collide and coalesce.) In fact, globular clusters are so ancient that they, along with the stars of the galactic hub, form a distinct 'stellar population' with its own chemical composition and life cycle (see page 342). They are thought to have formed during galactic mergers that created our own galaxy and are still shaping the evolution of others (see page 362).

Star death

All stars spend the vast majority of their lives steadily fusing hydrogen into helium in their cores, conforming to the well-understood laws of 'main sequence' stellar evolution (see page 270). But the last stages of the stellar life cycle, as the star exhausts its core hydrogen supply and must generate energy elsewhere in order to keep shining, see stars behave in very different ways depending on their mass.

All stars brighten and swell into red giants as they begin to burn fuel in the region around their core, but how long this can be sustained (and what happens next) depends crucially on their mass. For a star like the Sun, the ability to fuse the helium by-products of its hydrogen-burning phase offers a relatively brief respite, before the outer layers are shed in a short-lived but beautiful planetary nebula. The exhausted core, packed with the carbon, nitrogen and oxygen products of helium fusion, is exposed to space as a planet-sized, incandescent but slowly cooling white-dwarf star.

More massive stars have more spectacular ends – the heavier a star is, the longer it will be able to keep burning fuel in its core, fusing successively heavier waste products from the previous generation of fusion. Stars with more than eight times the mass of the Sun can eventually reach the natural end of this chain, creating a small but superdense core of iron. The fusion process cannot continue beyond this stage, and the abrupt 'switching off' of the core has spectacular consequences in the form of a supernova – a sudden inward collapse in the star's outer layers followed by a devastating explosion as the shock wave rebounds off the core.

Supernovae of this kind can briefly outshine an entire galaxy as the monster star consumes all of its remaining fuel in the course of a few weeks. They leave behind a glowing shroud of gas (enriched with heavy elements that can contribute to new generations of stars), and a collapsed stellar core that can take the form of either a city-sized neutron star or, if they have enough mass in their own right, a stellar-mass black hole (see pages 328 and 332). Such objects are most easily identified through their influence on other stars within binary star systems, which can sometimes produce dramatic effects (see page 334–9).

Red giants

Late in their lives, stars pass through one or more red-giant stages, in which their luminosity increases hugely and the pressure of radiation escaping from the interior causes the star's outer layers to balloon in size. As growth of the exposed surface area 'outpaces' the increase in luminosity, the star cools and reddens despite becoming vastly brighter.

The red-giant stage develops when a star exhausts its core supply of fuel (hydrogen in the first phase of its life, but later helium and even other heavier elements). As energy output from the core dwindles, the outer layers collapse until a thin shell around the core grows hot and dense enough to sustain nuclear fusion in its own right. In fact, the shell temperature can grow far hotter than the core, allowing nuclear fusion to proceed much more rapidly and boosting the star's luminosity, perhaps a thousandfold. Once fusion in the shell has begun, the energy it produces can sustain the reaction even as the star's upper layers are pushed out again by the increased radiation.

Antares
(red supergiant)

The Sun
(main-sequence
star)

Arcturus
(orange giant)

Temperature:
5,500°C
(9,900°F)

Temperature:
4,000°C
(7,200°F)

Temperature:
3,100°C
(5,600°F)

Red giants and supergiants are far
brighter than stars like the Sun, but
their greater size gives them much
lower surface temperatures.

Helium burning

As a star with the mass of the Sun or greater swells into a red giant, its core – exhausted of hydrogen and now packed with helium nuclei – contracts. Eventually temperatures reach such extremes (100 million °C or 200 million °F) that helium fusion can take place. A mechanism known as the 'triple-alpha process' combines three helium nuclei to make carbon, with a fourth sometimes joining to produce oxygen.

Helium fusion spreads rapidly through the core in an event known as the 'helium flash', whose consequences are initially hidden from view. Over time, however, the sudden increase in the core's luminosity causes the layers around it to expand and cool, slowing the rate of hydrogen fusion in the upper layers and, perhaps counterintuitively, causing the star to grow fainter and shrink in size. This stage in the life of a dying star, known as the 'asymptotic-giant' phase, is comparatively brief. When the core's helium is exhausted, shell burning resumes and the star swellls once again to a red giant.

In the 'triple-alpha' fusion process, a pair of helium nuclei (alpha particles) initially fuse to form beryllium (1), before the addition of a third particle creates carbon (2).

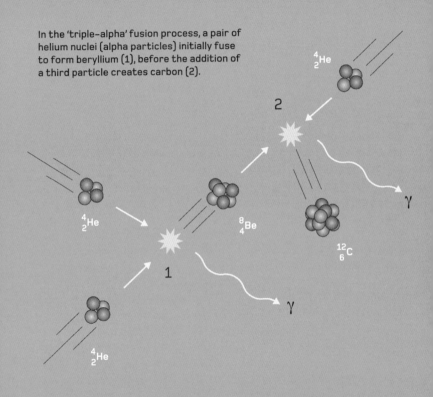

Planetary nebulae

Eventually, any Sun-like, low-mass star reaches a point at which it can no longer burn fuel of any kind in its core. The result is a star with an exhausted core, surrounded by one or more shells in which fusion continues. As these shells migrate outwards, the star grows ever more unstable, swelling and shrinking in size until eventually it puffs off its outer layers with such violence that they escape its gravity completely.

The resulting cosmic smoke ring, known as a planetary nebula, can take a variety of forms – most commonly a simple sphere or an hourglass, pinched in the middle by a denser ring of slower-moving material ejected in earlier times. The expanding material cools at first, but as the star's hot inner layers are exposed, it is energized by ultraviolet radiation that excites its atoms and causes it to glow in visible light. Planetary nebulae are short-lived phenomena in astronomical terms, shining for just a few tens of thousands of years before they dissipate, enriching the interstellar medium with their processed gases.

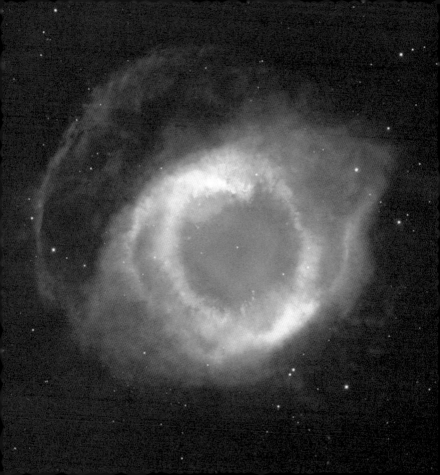

White dwarfs

A white dwarf is the exposed core of a Sun-like star, left behind after a dying red giant shrugs off its outer layers and nuclear fusion in the shells around its core falters and dies. With no energy to support the core from within, it collapses into a ball of helium, carbon, nitrogen and oxygen, roughly the size of Earth. As a result, despite surface temperatures that may start out as high as 200,000°C (360,000°F), white dwarfs are very dim and hard to detect. They usually give their presence away through their influence on other stars, as was the case with Sirius B, the faint companion of the brightest star in the sky. Over time, white dwarfs cool and fade, hypothetically becoming invisible 'black dwarfs'.

White dwarfs are 'degenerate' stars, prevented from further collapse by repulsive forces between the subatomic particles within them. This causes them to behave in curious ways – the more massive a white dwarf, the smaller it is, and this sets an upper limit on the mass of such objects (see page 328).

Sirius B (at lower left) is the faint white dwarf companion of the brightest star in the sky.

Supergiant stars

The most massive stars in the Universe are supergiants – heavyweights with the mass of tens of Suns, and luminosities hundreds of thousands, even millions of times that of the Sun. Despite the enormous pressure of radiation surging out from their cores, the gravity of these stellar monsters is high enough to keep them relatively compact, so that supergiants can display a variety of colours and form a broad horizontal band across the top of the Hertzsprung-Russell diagram (see page 268).

One consequence of the balance between gravity and radiation is that the more compact blue and white supergiants can develop particularly powerful stellar winds, allowing them to shed mass millions of times faster than more sedate Sun-like stars. They can shed entire solar masses of material in the course of their short lives, and as their outer layers are blown out into space, their hot interiors are exposed to space, forming a distinctive class of stars known as 'Wolf-Rayets'.

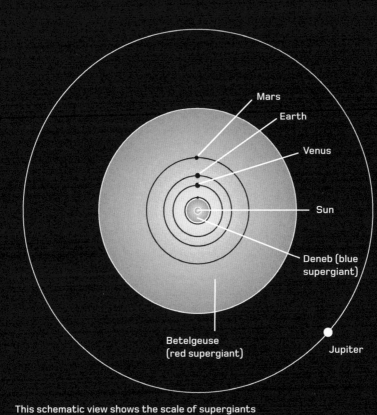

Mars

Earth

Venus

Sun

Deneb (blue
supergiant)

Betelgeuse
(red supergiant)

Jupiter

This schematic view shows the scale of supergiants
in comparison to our own solar system.

Supernova progenitors

Massive stars entering the final stages of their life become increasingly unstable and prone to spectacular explosions, which they survive more or less intact before a final cataclysmic detonation. The higher mass of supergiants and their kin allows them to sustain nuclear fusion of heavier elements than Sun-like stars can manage, transforming the products of helium fusion (carbon, nitrogen and oxygen) into heavier elements such as neon, sulfur and iron. But each new fusion reaction in the core is less efficient than the last, and runs for less time before it too spreads out beyond the core. The result is a star with a deep outer envelope of hydrogen surrounding a series of onion-like fusion layers and an increasingly dense core.

Perhaps the most famous and intensively studied supernova progenitor is Eta Carinae, a monster binary system in the constellation of Carina (see page 232), which last erupted in a brilliant 'supernova impostor event' in the 1840s.

The blue supergiants of Eta Carinae are surrounded by a double-lobed nebula of material that has been thrown off as they have become increasingly unstabla.

Type II supernovae

The sudden violent death of a giant star is known as a 'Type II' supernova, in contrast to other supernova phenomena (see page 338). They develop from a progenitor star whose onion-like layers of nuclear fusion have created a small but extremely dense iron core (see page 322).

This grows ever denser and hotter until it reaches the extremes required to fuse iron into heavier elements – but iron's nuclear structure means that it is the first element for which fusion actually *absorbs* energy, rather than releasing it. The outward pressure of radiation supporting the star is abruptly cut off, causing its upper layers to collapse under their own enormous gravity. The resulting shock wave compresses the core before rebounding to tear the star apart. The pressures and temperatures involved are so immense that they ignite nuclear fusion in the star's outer layers, allowing it to burn through tens of solar masses of fuel in a matter of days and creating a stellar beacon capable of briefly outshining an entire galaxy.

Stages of a Type II Supernova

1 Massive star builds up layered interior

2 Failure of fusion triggers core collapse

3 Shock wave rebounds from core and tears star apart in a fusion firestorm.

Supernova remnants

Supernovae play a key role in the evolution of the interstellar medium and the Universe as a whole, by seeding the space between the stars with heavy elements. Only stars considerably more massive than the Sun can produce elements heavier than carbon, nitrogen and oxygen, while those heavier than iron can only be forged in supernova explosions, where the vast amounts of energy available permit fusion processes that absorb energy rather than release it.

Supernova remnants such as the famous Crab Nebula in Taurus (aftermath of an explosion seen on Earth in AD 1054) are the shredded remnants of stars. They are visible for a short period due to their own intense temperatures (often in the tens of thousands of degrees), or via the heating effect as their expanding shock waves slam into the surrounding interstellar gas. They not only scatter heavy elements through space, but also trigger the birth of new stars as the passing shock wave compresses already dense clouds of gas and dust.

Neutron stars

The initial collapse involved in a Type II supernova (see page 324) puts the stellar core under immense pressure, triggering a collapse that creates the densest object in the Universe. The 'degenerate electron pressure' that halts the collapse of a planet-sized white-dwarf star (see page 318) is not strong enough to withstand this pressure and, as a result, the collapse continues unabated. As the atomic nuclei are forced together, they disintegrate into their constituent proton and neutron particles, and protons then combine with available electrons to create more uncharged neutrons. Provided the core's mass is not above a certain threshold (estimated at anywhere between 1.5 and 3.0 solar masses), the collapse eventually halts due to the build-up of 'degenerate neutron pressure'. By this time, the star will have dwindled to the size of a city, perhaps just 10 kilometres (6 miles) across. Most neutron stars only become visible if their magnetic field creates a pulsar, or if they interact with a companion star to form an X-ray binary system (see pages 330 and 334).

An X-ray view of the Crab Pulsar, the neutron star embedded in the heart of the Crab Nebula in Taurus (pictured in visible light on page 327).

Pulsars

When a stellar core collapses during a supernova explosion it retains most of its magnetism, concentrated by compression into a much smaller space. In fact, the magnetic field around a neutron star can be so intense that it channels the intense mix of radiations and particles blasting off the star's surface into two narrow beams of radio waves, visible light and X-rays aligned with its poles. Pulsars (pulsating stars) arise because the star's magnetic poles are rarely in perfect alignment with its axis of rotation, so they tend to sweep around the sky like lighthouse beams and sometimes happen to line up with Earth. Furthermore, thanks to its increased density, the neutron star ends up spinning much faster, so that pulsar beams tend to flash on and off in fractions of a second – small wonder that when the first pulsar was discovered in 1967, astronomers briefly wondered whether it might be an artificial signal from an alien civilization. Despite its power, the pulsar mechanism 'cuts out' after a few tens of millions of years as the star loses its rotational energy.

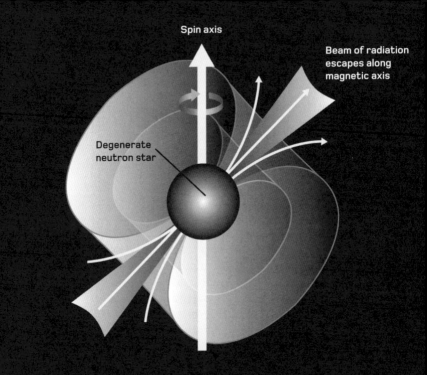

Spin axis

Beam of radiation
escapes along
magnetic axis

Degenerate
neutron star

A pulsar, or pulsating star, is a degenerate neutron
star – a collapsed stellar core with a mass of more
than 1.5 Suns – that beams radiation across space
like a cosmic lighthouse.

Stellar-mass black holes

What happens when a collapsing stellar core is so massive that its gravity can overcome even the powerful 'degenerate neutron pressure' that supports a neutron star (see page 328). The result is remarkable – the star's subatomic particles disintegrate even further into fundamental particles called quarks, and dwindle to a tiny point in space with infinite density, known as a singularity. This point has such strong gravity that nothing, not even light, can escape – it becomes effectively immune to direct observation, shielded behind a boundary called the event horizon (where the escape velocity exceeds the speed of light). This is a stellar-mass black hole, one of the strangest objects in the Universe: it will pull in anything that strays too close, growing in both mass and the extent of its event horizon as it feeds. Such objects were hypothesized as long ago as the 1780s, but it is only since the 1930s that we have had a theoretical model to explain their formation, and since the 1970s that we have been able to identify 'candidate' black holes through their influence on neighbouring stars.

Ergosphere – region of distorted, rotating spacetime

Event horizon – 'point of no return' where escape velocity > c

Flat spacetime at a safe distance from black hole

'Gravitational well' of distorted spacetime

Singularity

Extreme binaries

When two stars in a binary system have significantly different masses, the heavier one may race through its life cycle quite rapidly, while its less massive companion remains in sedate middle age. As a result, the lower-mass star may end up in a binary relationship with a dense stellar remnant. Measuring the behaviour of the normal stars in such systems is a useful technique for 'weighing' stellar remnants, but the intense gravity field around the remnant can also give rise to a variety of extreme phenomena – especially as the secondary star nears the end of its life and swells to become a red giant.

The giant star's tenuous outer envelope is particularly vulnerable to being distorted or pulled away by its neighbour, creating an 'accretion disc' as it spirals down onto the stellar remnant. In the case of white-dwarf remnants, the result may be a nova system (see page 336). The tidal forces of a neutron star or black hole, meanwhile, may heat the disc to millions of degrees, creating a so-called X-ray binary system.

This artist's impression shows a dense stellar remnant pulling material away from the atmosphere of its larger but less massive companion.

Nova systems

A nova is a binary star system in which the higher-mass component has already evolved to become a white dwarf, while the originally less-massive component has now become a red giant. As the red giant swells massively in size, the white dwarf tugs at its hydrogen-rich outer layers, creating a stream of gas that spirals down onto the surface of the dense stellar remnant and builds up into a hot 'atmosphere', compressed to enormous density by the dwarf's powerful surface gravity. Eventually, conditions in the atmosphere may become so extreme that they mirror those in the heart of a Sun-like star, triggering a huge burst of nuclear fusion that burns through the accumulated gas in a matter of weeks.

Nova eruptions can see the star system's overall light output multiply ten-thousandfold, and may be one-off events triggered during a very brief phase of the system's evolution, or a longer-lived recurrent phenomenon, with eruptions every few decades or more.

Companion star

White dwarf

Atmosphere builds
up on outer layers

Material pulled away
from companion

Nova event burns off
atmosphere in a brief
burst of nuclear fusion

Type Ia supernovae

While most supernova explosions mark the sudden death and core collapse of a massive star (see page 324), Type Ia supernovae are a special class that have a very different cause. They occur in nova systems (see page 336) whose white-dwarf component is close to the upper 'Chandrasekhar limit' for white dwarfs of about 1.4 solar masses (see page 328).

If the white dwarf accumulates enough extra material to tip over this limit, the results are spectacular. As its gravity overcomes the 'electron degeneracy pressure' that has hitherto supported it from within, a runaway nuclear fusion reaction begins, burning through the white dwarf's carbon in a matter of seconds, and triggering a supernova explosion. Because this kind of supernova process always involves a star of a very specific mass, it always liberates the same amount of energy and reaches the same peak luminosity. This makes Type Ia supernovae in distant galaxies valuable 'standard candles' for measuring the scale of the Universe (see page 402).

Supernova 1994D was a Type Ia supernova seen on the outskirts of galaxy NGC 4526 in 1994. Visible at lower left, it was almost as bright as its entire host galaxy, despite a distance of 108 million light years.

Galaxies

Ancient Greek astronomers saw the broad band of pale light that runs across the night sky as a stream of spilt milk, the *Via Galactea* or Milky Way from which the modern word 'galaxy' derives. Today, we understand that the Milky Way in the sky is a geometrical effect created by our solar system's location in the outlying regions of a huge, flattened disc of stars. When we look in certain directions, we look along the plane of the galaxy, towards distant star clouds that stretch away for thousands of light years. Looking in other directions, we see through the galactic plane, past the relatively scattered stars in our region of the galaxy, and out into the vastness of intergalactic space.

The curious spirals, clouds and balls of stars found in these parts of the sky were only identified as independent galaxies in their own right in the early 20th century, thanks largely to Edwin Hubble's use of Cepheid variable stars (see page 348) to determine their distances. Hubble also went on to discover the general expansion of the Universe (see page 384), and also drew up the first classification of different galaxy types, recognizing distinct groups known as spirals, barred spirals, ellipticals, lenticulars and irregulars (thanks to Herculean efforts to map the distribution of stars and gas in our own galaxy, we now know that the Milky Way is a barred spiral).

Hubble's basic scheme is still used today, often illustrated with a 'tuning fork' diagram that he devised (see page 357). But other galaxies soon emerged that refused to fit his scheme and were therefore classified as 'peculiar'. Most of these have since proved to be the result of collisions, mergers or close encounters between galaxies that can individually be fitted into the Hubble scheme: galaxy interactions have proved to be far more common than predicted, thanks to their relatively crowded distribution and gravitational attraction to each other. Often, they trigger bursts of activity from the region of the galactic nucleus, creating an unusually bright 'active galaxy' (see page 366). While Hubble thought that the variety of galaxies showed a simple evolutionary progression from ellipticals to spirals (implicity in his tuning fork diagram), later research has effectively turned this idea on its head, revealing that here too, galaxy collisions play a crucial role (see page 372).

The Milky Way

According to the latest evidence, our Milky Way galaxy is a huge spiral structure with a visible disc of stars roughly 100,000 light years across. The central hub itself is a huge ball about 20,000 light years in diameter, with a bar of stars emerging from either end. The entire visible galaxy, meanwhile, is surrounded by a large 'halo' region – this is where globular clusters (see page 308) are mostly found, and is also home to large quantities of unseen 'dark matter' that affect the galaxy's overall rotation (see page 400) and account for perhaps 90 per cent of its mass.

The Milky Way contains an estimated 400 billion stars, with significant differences in terms of both age and chemical composition between different regions. The disc is dominated by relatively young, metal-rich 'Population I' stars, with the youngest and shortest-lived concentrated in the spiral arms. The hub and globular clusters, meanwhile, contain mostly small and ancient red and yellow stars known as 'Population II' stars.

Features of the Milky Way galaxy

1 Galactic disc
2 Galactic bulge
3 Halo containing globular
  star clusters

4 Supermassive black hole
5 Position of solar system

Spiral arms

The spiral arms that whirl their way out from the central hub of our galaxy and many others are simply the brightest areas within an overall disc that is crowded with stars. They owe their prominence to open clusters of brilliant young stars, as well as luminous pink-hued emission nebulae (see pages 306 and 276). However, they are not coherent objects – if that were the case the arms would wind themselves onto the galaxy's hub in the course of a few rotations.

Bright, massive stars with short lifespans live and die close to their birthplace, demonstrating that the spiral arms are regions of concentrated star formation. Astronomers think that the arms are in fact 'density waves' – galactic traffic jams where objects orbiting the hub slow down and become more densely packed together, triggering the process of starbirth and the formation of the bright clusters that highlight their presence. The waves may be caused by the gravity of other nearby galaxies affecting individual orbits within the disc.

A simplified overhead view
of the Milky Way galaxy

Scutum–Centaurus Arm

Sagittarius Arm

Norma Arm

Central bar

Perseus Arm

Orion Spur

Outer Arm

The galactic hub

The central region of our galaxy is a huge ball of stars concentrated in a region 20,000 light years across. It is dominated by ancient red and yellow 'Population II' stars, but the regions closest to the galactic centre, 26,000 light years away, contain open star clusters with some of the most massive stars known. Swirling superhot gas is also visible at X-ray wavelengths through the intervening star clouds.

The turbulence around the galactic centre is caused by the secret at the heart of the Milky Way – an enormous 'supermassive' black hole with the estimated mass of 4.5 million Suns. Astronomers now believe that most galaxies have similar monsters at their centres, created during their formation and sometimes producing spectacular activity as they feed on stars, gas and dust in their surroundings (see page 366). Although the Milky Way's black hole is effectively dormant, it can be measured through its gravitational influence on nearby stars that orbit it with incredible speed.

This artist's impression shows the orbits of stars and a recently discovered gas cloud close to the unseen Sagittarius A* black hole.

The distance to galaxies

Galaxies beyond our own are so remote that it's impossible to measure their distances using conventional techniques such as parallax (see page 258). Instead, for relatively nearby galaxies, astronomers look for Cepheid variable stars – yellow supergiants whose cycle of pulsations is related to their average luminosity. The longer a Cepheid's period, the more luminous it is, so by measuring the cycle of a distant Cepheid and comparing its true luminosity with its brightness from Earth, it is possible to estimate its distance. This kind of approach, using an object of known luminosity to measure an unknown distance, is known as a 'standard candle' technique.

Beyond our cosmic neighbourhood, however, astronomers must rely on other techniques. Rare Type Ia supernovae (see page 338) can be used as standard candles over much greater distances (see page 402), and thanks to the overall expansion of the Universe, a galaxy's red shift can be taken as a rough indication of its distance from Earth (see page 384).

Hubble's Law relates a galaxy's 'recessional velocity' to its distance from Earth, showing that they are linked by the Hubble Constant – roughly 21.5 km per second (13.5 miles per second) per million light years.

Velocity away from Earth (indicated by red shift)

Gradient of line = Hubble constant (velocity/distance)

Distance from Earth

Satellite galaxies

O ur Milky Way's powerful gravity traps well over a dozen smaller star clouds in orbit around it. The largest and most famous of these are the Large and Small Magellanic Clouds, which lie about 160,000 and 200,000 light years away from Earth respectively. The two galaxies appear as isolated clumps of stars in the southern constellations of Dordado and Tucana (see pages 246 and 252), and both follow more or less identical billion-year orbits around our own galaxy, leaving behind a faint trail of stars and gas known as the Magellanic Stream. Significantly closer to Earth, but much smaller, are the Canis Major Dwarf and Sagittarius Dwarf Spheroidal Galaxies, roughly 25,000 and 70,000 light years away. These faint star clouds were detected in 1994 and 2003 as unexpected 'overdensities' in the distribution of stars — their orbits suggest they have passed directly through the Milky Way in the past and will do so again in the future. Such collisions strip the smaller galaxies of their stars, but may also help to reinforce the Milky Way's spiral structure (see page 344).

A projected view of the sky seen from Earth shows
the plane of the Milky Way across the centre, with the
Magellanic Clouds and their associated stream of
radio-emitting gas to the south.

The Local Group

Galaxies are gregarious – the distribution of material out of which they formed, coupled with their own substantial gravity, gives them a tendency to form clusters (see pages 370 and 374). Our own Milky Way spiral is one large member of a relatively small association called the Local Group, containing at least 54 galaxies in a region 10 million light years across. The other major components are the Andromeda Galaxy (Messier 31 – see page 160) and the Triangulum Galaxy, (M33 – see page 166). M31 is a barred spiral that is both larger and brighter than the Milky Way (though less massive), while M33 is a smaller and less well-defined spiral thought to be in orbit around M31.

Both the Milky Way and M31 have a substantial number of smaller satellite galaxies in orbit around them, with other small galaxies also included as members of the overall group. The two major galaxies are also falling towards one another thanks to their mutual gravitational attraction and are doomed to collide and eventually merge around 5 billion years from now.

Spiral galaxies

Beautiful pinwheels in space, spirals are perhaps the archetypal galaxies, accounting for 60 per cent of star systems in the nearby Universe. They show considerable variation in size: at 100,000 light years across, the Milky Way is perhaps three times the size of the smallest spirals.

Spirals are broadly divided into two groups – normal spirals (class 'S') in which the arms emerge directly from the hub region, and barred spirals (class 'SB') in which the arms are rooted at the ends of a straight 'bar' that runs across the hub. The spiral arms are just the brightest and most prominent regions in a flattened disc of stars orbiting the hub, and may vary considerably in brightness and definition. 'Grand design' spirals have very well-defined arms, probably created by interaction with neighbouring galaxies (see page 362), while 'flocculent' spirals have much looser clumps of star formation, probably arising when the large-scale spiral pattern is weak enough to be overwhelmed by local influences.

Elliptical galaxies

Huge balls of ancient red and yellow stars, elliptical galaxies include both the largest and smallest star systems, and account for about 20 per cent of all galaxies. They vary in shape from perfect spheres to elongated rugby balls and even cigar shapes (classed from 'E0' to 'E7' in order of increasing elongation). The smallest 'dwarf elliptical' or 'dwarf spheroidal' galaxies may be just a few thousand light years across and can vary hugely in density from tens of thousands of stars to many billons. The largest 'giant ellipticals' are the biggest galaxies in the Universe, containing trillions of stars.

In stark contrast to spiral galaxies, ellipticals contain very little interstellar gas and dust – hence, they are dominated by small, long-lived 'Population II' stars that have shone since ancient times when the galaxies were richer in star-forming material. Individual stars follow elliptical orbits around the galaxy's central regions at a wide variety of inclinations, and the overlap of countless orbits gives the galaxy its shape.

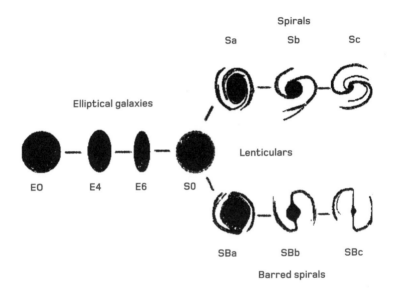

Spirals

Sa Sb Sc

Elliptical galaxies

Lenticulars

EO E4 E6 SO

SBa SBb SBc

Barred spirals

Edwin Hubble's famous galaxy classification scheme identifies spirals (S) and barred spirals (SB) by the tightness of their arms and ellipticals (E) by their shape, from spherical EO to elongated E6.

Irregular galaxies

The third major group of galaxies, alongside the spirals and ellipticals, are the irregulars (catalogued 'Irr'). They make up about a quarter of all galaxies in the present-day Universe, though they appear to have been much more abundant in earlier times (see page 380).

As their name suggests, irregulars are more or less shapeless clouds of stars, gas and dust, though they are usually separated into two distinct categories. Irr-II galaxies are entirely lacking in structure, while Irr-I galaxies – such as the Milky Way's biggest satellite, the Large Magellanic Cloud – show signs of incipient spiral features such as arms and central bars. They tend to be smaller than most spirals, but can be bright for their size, since they are often rich in sites of ongoing star formation and short-lived but brilliant heavyweight stars. In the most spectacular irregular 'starburst' galaxies, star formation is taking place at a vastly accelerated rate compared to normal galaxies like our own.

NGC 1569 is an irregular galaxy in Camelopardalis
that is currently undergoing a huge starburst
as it interacts with other nearby galaxies.

Lenticular galaxies

A fairly rare intermediate class of galaxy, lenticulars have a large hub of old red and yellow stars, a flattened disc of middle-aged stars and dust, but very little interstellar gas. They therefore resemble either an elliptical galaxy with a surrounding stellar disc, or a spiral galaxy without the spiral arms (and indeed, they are classified as 'S0' galaxies). A few show extended stellar bars emerging from their hubs, or even ghostly 'spokes' that may be precursors of spiral arms.

There are two competing theories to explain the origin of lenticulars, neither of which is entirely satisfactory. One is that they are simply old spirals that have used up most of their gas and therefore stopped forming new stars in spiral arms. However, this cannot explain the unusual size of their hubs. The other suggests that they form from galactic mergers that drive away their interstellar gas (see page 372) swelling the size of their hubs while bringing a premature halt to star formation. This, however, cannot explain the survival of the disc structure.

Galaxy collisions

Although the hundreds of thousands of light years that separate large galaxies are vast distances by any normal scale of measurement, galaxies are actually rather crowded together when considered on their own terms. As a result, collisions between them are far more common than they are between, for example, stars or planets. As telescope technology has improved over the past few decades it has become clear that many of the strange-looking 'peculiar' galaxies that did not previously fit into any known classification are actually the result of galaxy collisions and mergers. Many more galaxies are affected by strong tides raised during close encounters with their neighbours.

Collisions and interactions are thought to play an important role in the evolution of galaxies (see page 372). They also seem to drive the development of specific features such as active galactic nuclei (see page 366), and even strong 'grand-design' spiral structures.

Giant ellipticals

The biggest galaxies in the Universe, giant ellipticals are enormous balls of stars up to a quarter of a million light years across. Comparatively rare, they are found only in the centre of large galaxy clusters (see page 370). The largest of all, so-called 'cD-type' galaxies, may have diffuse halos of stars up to 3 million light years across. Though fairly free of star-forming gas, and therefore dominated by long-lived red and yellow stars, they often display unusual activity around their cores (see page 366), appearing as strong sources of both radio waves and X-rays.

Giant ellipticals are thought to be created by the collision and merger of many smaller galaxies in the crowded central regions of galaxy clusters – as their mass increases, so does their gravitational influence. They grow to enormous size by cannibalizing their neighbours in collisions that send stars into chaotic elliptical orbits and feed material to their enormous central supermassive black holes.

Active galaxies

While most galaxies shine with the combined light of billions of individual stars, some show signs of other processes at work. Otherwise-normal spiral galaxies sometimes display brilliant, starlike sources of light in their cores, and are known as Seyfert galaxies. Radio galaxies are surrounded by billowing clouds of radio-emitting gas, spanning hundreds of thousands of light years and often connected to the core by narrow jets. Quasars and blazars, meanwhile, are starlike in appearance and vary their radiation output from day to day. At first they were believed to be unusual 'radio stars', but today it's clear that they are actually the astoundingly bright central regions of some of the most remote galaxies in the Universe.

Astronomers believe that all these active galaxies are manifestations of the same phenomenon, an active galactic nucleus (see page 368). The exact appearance of a galaxy depends on the amount of activity occuring in its nucleus and the angle at which the system presents itself to Earth.

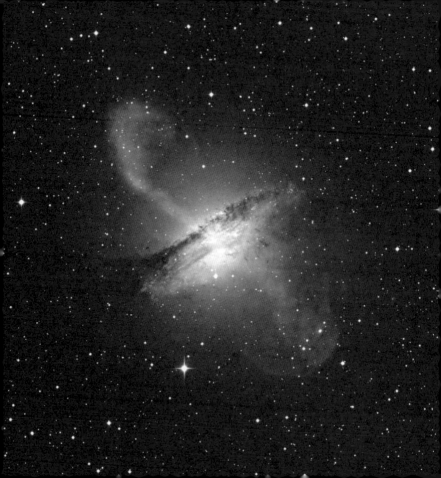

Active galactic nuclei

Intensely luminous and violent sources of both radiation and high-energy particles, active galactic nuclei are believed to be caused by the supermassive black holes at the heart of distant galaxies feeding voraciously on gas, dust and stars from their immediate surroundings. As material is drawn in by the gravity of the solar-system-sized black hole, it is torn apart by intense tidal forces to create an 'accretion disc'. Heated to millions of degrees, it becomes an intense source of visible light and high-energy radiations. This is the central light source in quasars, blazars and the less active Seyfert galaxies, but it is often surrounded by a dense doughnut-shaped cloud of dust and gas that hides it from view when the galaxy is seen edge on. Tangled magnetic fields around the central black hole, meanwhile, sweep up some of the infalling material and spit it out in high-speed jets along the galaxy's axis of rotation. As these particles slam into nearby intergalactic gas clouds, they billow out into the enormous radio-wave-emitting clouds characteristic of radio galaxies.

1 Surrounding 'host' galaxy

2 Central black hole draws in material

3 Accretion disc of superhot material
  around black hole forms quasar

4 Jets of material escape above
  and below quasar

5 High-energy particles in jets
  emit bright radiation

Galaxy clusters

Clusters of galaxies range from the relatively sparse, like our own Local Group (see page 352), to crowded regions in distant space. Small groups contain a few dozen dwarf galaxies and perhaps two or three major members, while larger full-blown clusters may contain dozens of big galaxies and hundreds of accompanying dwarfs. Remarkably, however, clusters usually occupy a similar volume of space (perhaps 10 or 20 million light years across) regardless of their density.

Visible galaxies account for only 1 per cent of a cluster's mass – X-ray studies typically reveal ten times as much sparsely scattered, superhot gas, concentrated towards the centre of a cluster. The gas is believed to be heated and driven off during the galaxy mergers that give rise to elliptical galaxies, becoming energetic enough to escape the gravity of individual galaxies while remaining bound to the cluster as a whole. The vast majority of mass in a cluster, however, takes the form of unseen 'dark matter' (see page 400).

Galaxy evolution

Many aspects of the story are still uncertain, but most astronomers today believe that collisions drive the evolutiuon of galaxies from one type to another, with interstellar gas playing a key role. Gas clouds, though sparse, exert strong gravitational pull on other objects such as stars: they are also much more likely to collide and heat up during galactic mergers, even as stars pass each other unscathed.

This helps build a picture of the stages involved in galaxy building. Spirals, it seems, formed from the merger of abundant small irregular galaxies in the early Universe (see page 380). Large-scale collisions between spirals in the modern Universe, in contrast, trigger short-lived bursts of star formation, but ultimately heat the gas within them enough for it to escape into intergalactic space. With little or no gas to sustain new star formation or guide stellar orbits, the end result of multiple spiral mergers is a chaotic ball of stars in overlapping random orbits – an elliptical or giant elliptical galaxy.

Superclusters

Galaxy clusters are rarely isolated in space – instead, they tend to merge together at their edges, creating enormous superclusters that may be hundreds of millions of light years across. Our own Local Group is an outlying member of a Local Supercluster centred on the dense Virgo Cluster, about 50 million light years from Earth. The gravitational attraction of the central massive cluster is enough to overcome the general expansion of the cosmos (see page 384), causing superclusters to grow denser over time – for example, gravity is drawing us towards the Virgo Cluster at a speed of about 5.4 million kilometres per hour (3.4 million mph).

Superclusters in their turn blur together at their outer boundaries, linking up with one another to form huge chains and sheets known as 'filaments' surrounding even more enormous, apparently empty 'voids'. These are the largest structures in the cosmos – remnants of variations in the density of the very early primordial Universe (see page 392).

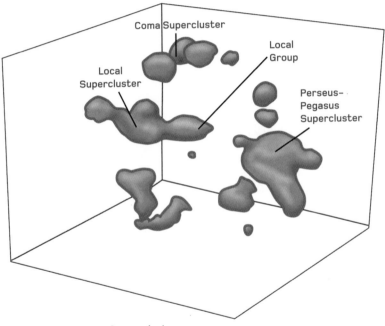

Coma Supercluster

Local
Group

Local
Supercluster

Perseus-
Pegasus
Supercluster

A schematic map of superclusters
in the nearby Universe.

Cosmology

Cosmology is the branch of astronomy that looks at the properties of the Universe as a whole – its fabric, structure on the largest scale, origins and possible fates. For most of human history, many of the questions raised by cosmology were simply not on the agenda – the Universe was seen as a small and relatively recent creation. As the direct result of supernatural intervention by a divine intelligence, it was not something that needed to be explained through the kind of physical laws that govern the cosmos today.

It was only in the 19th century that breakthroughs in geology and physics began to suggest that the Earth was millions of years older than any creation story suggested. This came with a realization that its features had built up largely through the slow, inexorable action of the same processes shaping it today, over 'geological' timescales. The discovery that the Earth was so much older than anyone had suspected forced a radical rethink in the understanding of the wider Universe, and many astronomers went to the opposite extreme, proposing a Universe of infinite age, with no beginning and no end.

The key evidence that forms the backbone of modern cosmology emerged in the 1920s. Many astronomers laid the foundations, but the ultimate synthesis was the work of US astronomer Edwin Hubble. It was he who first conclusively proved the existence of galaxies beyond our own, and who then went on to prove that the Universe as a whole is expanding (see pages 348 and 384). This inspired theoreticians such as Belgian priest Georges Lemaître to think about conditions in the distant past, when the Universe would have been smaller, denser and hotter than it is today.

It was Lemaître who first proposed a cosmic origin in the sudden expansion of what he called a 'primeval atom' – the concept known today as the Big Bang – but it took a combination of relativity, quantum and atomic physics to reveal how this fiery birth could give rise to matter in the forms we know today. Yet, while the evidence suggests that the Big Bang theory is correct in its broad outlines, the precise details of the origin and fate of the cosmos are still a subject of intensive research.

The large-scale Universe

The largest structures in the Universe are known as filaments and voids. Filaments are huge chains and sheets formed by galaxy superclusters stretching across hundreds of millions of light years (see page 374), while voids are the even larger, and apparently empty, spaces in between them. Maps of galaxy distribution suggest that the Universe has a 'Swiss cheese' appearance – distinctly uneven on the relatively local scale, but essentially uniform across the entire span of the cosmos.

While gravitational collapse is responsible for the individual galaxies found within the filaments, the scale of these enormous structures is far too large to be explained solely by the actions of gravity. Instead, filaments and voids are thought to trace their origins back to variations in the density of the primordial Universe that were already in place shortly after the Big Bang and whose traces are preserved in the Cosmic Microwave Background Radiation (see page 382).

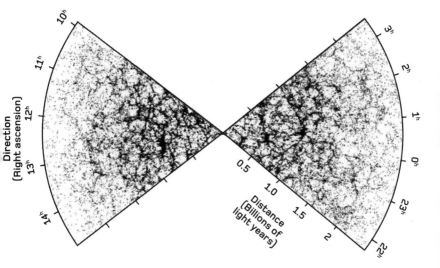

A map of the nearby Universe, produced from an
Anglo–Australian Observatory galaxy survey,
reveals a cosmic web of filaments.

The Hubble Deep Fields

The deepest views of the Universe so far available come from the unblinking eye of the Hubble Space Telescope in orbit above the Earth. A series of 'Hubble Deep Field' images have revealed galaxies stretching away across billions of light years of distant space. Because their light has taken so long to reach us, we are seeing these galaxies as they were when the cosmos was much younger – over these huge distances the Universe acts like a vast time machine enabling us to see back into the distant past. The most distant galaxies of all are faint red blobs, their light amplified by the 'gravitational lensing' effect of dense galaxy clusters closer to us (see page 386).

The population of galaxies revealed in the Deep Field images is very different from those found in the modern Universe: bright irregular galaxies are common, spirals less so, and ellipticals all but absent, while the violent active galaxies known as quasars are also widespread. These differences are important evidence for galaxy evolution through repeated mergers (see page 372).

A wide variety of galaxies – often distorted by interactions with their neighbours – are visible in this small section of the 'Hubble Ultra-Deep Field'.

Cosmic Microwave Background Radiation

While the most distant parts of the sky appear pitch-black in visible light, in 1964 US astronomers Arno Penzias and Robert Wilson discovered that they are in fact glowing softly at radio wavelengths. This soft glow of radiation, seen in all parts of the sky, is the afterglow of the Big Bang itself, heating the Universe to a residual temperature of 2.73°C (4.91°F) above the coldest possible temperature, absolute zero.

The Cosmic Microwave Background Radiation (CMBR) was released as the expanding Universe became transparent roughly 380,000 years after the Big Bang (see page 396). At the time, it was emitted as mostly visible radiation, but during its long journey to reach us, the expansion of space has robbed it of energy, reducing it to low-energy radio waves. Nevertheless, the CMBR offers us a unique look back into conditions shortly after the Big Bang: tiny variations in its temperature and density show the seeds of large-scale cosmic structure, sown during the period known as Inflation (see page 392).

A map of the CMBR from NASA's Wilkinson Microwave Anisotropy Probe (WMAP) reveals tiny fluctuations between hotter areas of the primordial Universe (bright spots) and cooler ones (dark regions).

Cosmic expansion

The development of a method to measure the distance to other galaxies in the early 20th century was soon followed by an even greater breakthrough. Astronomers had long recognized that the spectra of 'spiral nebulae' revealed very large red shifts in their light, suggesting that (if the red shifts were due to the Doppler effect – see page 264) they were moving away from Earth at high speeds. When Edwin Hubble compared his measurements of galaxy distances with their red shifts, a remarkable pattern emerged – on average, the further away a galaxy is, the faster it is moving away from us.

The implications of this relationship, known as Hubble's Law, are immense: they do not mean that our galaxy is uniquely unpopular, but rather that space itself is expanding, dragging all galaxies away from each other like currants in a rising cake. The constant linking distance and speed of recession, known as the Hubble Constant, is currently measured at 21.5 kilometres per second (13.3 miles per second) per million light years.

The expanding Universe carries galaxies away from each other as if they were stuck on the surface of a balloon. Because all of space is expanding at the same rate, more widely separated galaxies move apart with greater speed.

More widely spaced galaxies in the future

Universe today

The distant past, when galaxies were closely packed

The nature of spacetime

Even before the discovery of cosmic expansion and the development of the Big Bang theory of cosmic origins, the early 20th century saw a revolution in our understanding of space and time thanks to the work of Albert Einstein. His 1905 theory of special relativity rewrote the laws of physics to take account of the fixed speed of light, c (300,000 kilometres per second or 186,000 miles per second), revealing consequences such as the inconstancy of space and time measurements, and the fact that mass and energy are equivalent and exchangeable according to the famous equation $E = mc^2$.

His 1915 paper on general relativity, however, went further than this, explaining that gravity itself is a distortion in the fabric of a four-dimensional 'spacetime', caused by the presence of mass. Einstein's theories have little effect on our experience of the everyday world, where the laws of classical physics still apply, but astronomers frequently see their effects when they look at situations involving extreme motion or intense gravity.

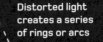

Distorted light
creates a series
of rings or arcs

Distant
galaxy

Intervening galaxy
cluster

Cluster's gravity
bends path of light
from distant galaxy

Observer on Earth

Fundamental forces and the quantum Universe

While gravity is the primary force shaping the large-scale Universe, it is just one of four known fundamental forces. The others are electromagnetism (which makes its presence felt through phenomena such as electricity, light and magnetic fields), the strong nuclear force (which binds particles together on the tiny scales of the atomic nucleus) and the weak nuclear force (responsible for effects such as radioactive decay).

Quantum physics governs matter and energy on the smallest scales, but this has some important implications on the larger scales of cosmology, too. One important issue is that quantum-scale phenomena are 'fuzzy' – prone to random variations and best described in terms of probabilities rather than the mechanistic certainties of classical physics. Many cosmological observations, such as the fluctuations in the CMBR and the very distribution of matter itself can trace their origins to quantum effects in the early Universe (see pages 382 and 392).

According to current theories, the present-day fundamental forces can be unified as 'superforces' under extremely high energy conditions, and may ultimately have originated as a single force, a so-called 'Theory of Everything', in the Big Bang itself.

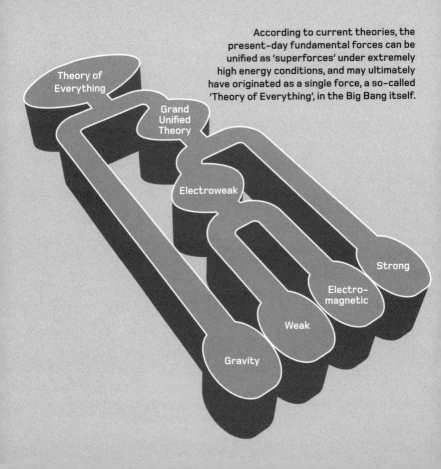

The Big Bang

The Big Bang theory has been our best model of cosmic origins for more than half a century. Based on evidence such as cosmic expansion and the CMBR (see pages 384 and 382), and informed by quantum physics and general relativity (see pages 388 and 386), it suggests that the Universe came into being in a hot, dense state about 13.8 billion years ago.

The original Big Bang was a sudden burst of energy that ultimately created not only all the matter in the Universe, but also space and time themselves – making it nonsensical to talk about 'where' it happened or what happened 'before'. However, various trigger events have been proposed, including the collapse of a previous Universe, and a collision between many-dimensional structures called 'branes'. Quantum theory dictates that the laws of physics become entirely unpredictable when the cosmos was below a certain size, making it impossible to predict the behaviour of the Universe back further than 10^{-43} seconds after the Big Bang itself.

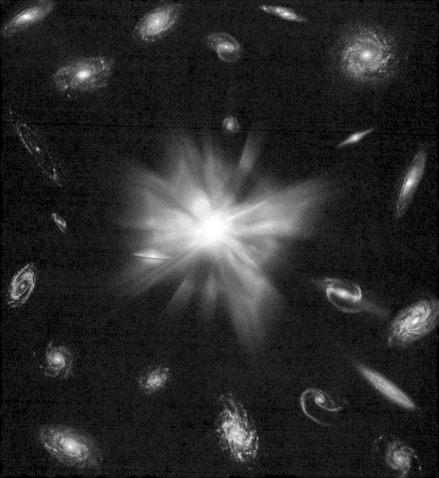

Inflation

A brief but vital episode in cosmic history, inflation saw a tiny region of the infant Universe grow rapidly between about 10^{-36} and 10^{-32} seconds after the Big Bang itself, producing a bubble that is the seed of our present-day cosmos. The event is thought to have been triggered by a burst of energy associated with the separation of an initial 'superforce' into the present-day fundamental forces (see page 388).

Inflation was first proposed by US physicist Alan Guth in order to explain the existence of large-scale structure in the cosmos. Without it, the Big Bang theory predicts a very smooth distribution of matter and little or no structure. Inflation 'magnified' tiny quantum fluctuations in the early Universe to much larger scales, allowing them to produce the variations in the density of matter that eventually gave rise to galaxy clusters and superclusters. The discovery and mapping of 'ripples' in the CMBR (see page 382) have confirmed Guth's idea as essentially correct.

The expanding
universe

1 Big Bang
2 Cosmic Inflation
3 Cosmic 'dark age' before
   stars begin to form
4 Temperature falls over time
5 Size of observable universe

The birth of matter

The Big Bang released vast amounts of energy, but how did it give rise to particles of matter, and eventually atoms themselves? The process involved the spontaneous creation and destruction of paired particles of matter and antimatter on very short timescales. These annihilate each other on contact to release energy, so the early Universe was alive with particles winking in and out of existence. Hotter conditions in the earliest times allowed relatively massive particles known as quarks and antiquarks to be created, alongside light particles called leptons and antileptons. For complex reasons, the balance tipped slightly in favour of quarks and leptons, so that as temperatures dropped below those needed to create the

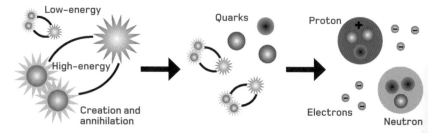

Low-energy

High-energy

Creation and annihilation

Quarks

Proton

Electrons

Neutron

various particles (after just a few seconds), a small amount of 'normal' matter survived.

As the Universe continued to expand and cool, particles began to come together in stable configurations. Triplets of quarks joined to form either positively charged protons or uncharged neutrons, some of which bonded in turn to form atomic nuclei of the lightest elements. After 20 minutes, this phase came to an end, leaving vast numbers of lone protons to form the nuclei of hydrogen, the simplest element of all. Over the next 380,000 years, the Universe continued to expand as a hot, opaque 'soup' of atomic nuclei and electrons (the dominant type of lepton). The final stage in the formation of matter only came when temperatures dropped low enough (below about 2,700°C or 4,900°F) for electrons to enter stable orbits around the nuclei – an event known as 'recombination' that would radically alter the Universe (see page 396).

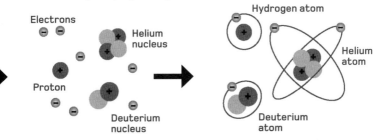

The cosmic dark age

The 'recombination' event that marked the final phase in the creation of matter, about 380,000 years after the Big Bang itself (see page 390) saw free-floating electrons 'soaked up' by atomic nuclei. This caused a dramatic drop in the density of particles and, as a result, the Universe became transparent for the first time. Previously, electromagnetic radiation had bounced around within a 'fog' of particles, constantly being absorbed and re-emitted, but now for the first time photons of light could travel in straight lines. The intimate link between matter and radiation was severed (an event known as 'decoupling'), and radiation released at this time is now seen all across the sky in the form of the CMBR (see page 382).

In the aftermath of decoupling, the Universe was suddenly dark for the first time. With no stars to illuminate it, this period is known as the 'cosmic dark age' — it lasted for about 400 million years, during which matter continued to slowly coalesce in the darkness, forming the seeds of the first stars and galaxies.

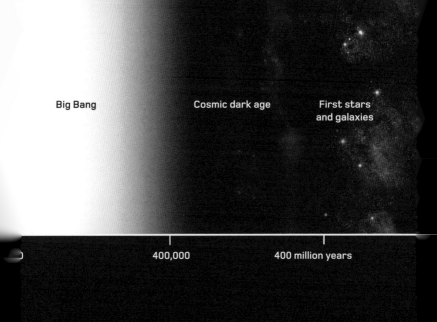

Big Bang

Cosmic dark age

First stars and galaxies

400,000

400 million years

This schematic depicts the clearing fog of the early Universe following recombination, the ensuing cosmic dark age, and the emergence of the first luminous stars and galaxies.

First stars and galaxies

About 13.3 billion years ago, the Universe began to emerge from its dark age, as collapsing knots of gas in the darkness grew dense enough to begin nuclear fusion and the first stars were born. Consisting only of light elements, these 'Population III' stars were immune to the CNO cycle (see page 274), and so could accumulate masses equivalent to several hundred Suns without blowing apart. Despite being extremely luminous, they lie beyond the reach of current telescopes, so much of what we know about them relies on computer models.

Their deaths after a few million years enriched the interstellar medium with heavier elements, beginning a process that is still continuing today. They also left behind monstrous black holes that began to draw in material from their surroundings at a tremendous rate. Irregular galaxies began to form around these seeds, growing rapidly through repeated mergers, while feeding more material to the central black hole to create the active galactic nucleus of a quasar (see page 368).

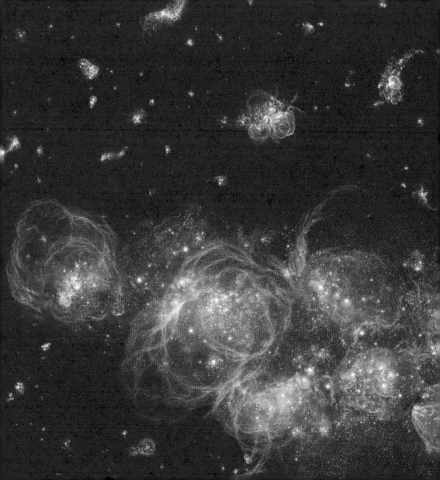

Dark matter

Since the late 20th century, astronomers have understood that the Universe's visible matter, from stars and gas to dust clouds and planets, is vastly outweighed by material that is not only invisible, but also transparent. Dark matter gives away its presence only through gravitational effects – it was first detected in the 1930s through the behaviour of galaxies orbiting within clusters, and confirmed in the 1970s through studies of the rotation of stars within the Milky Way.

Dark matter makes up an astonishing 84.5 per cent of all the mass in the Universe, but astronomers have little idea of what it is. One early theory that it consisted largely of small, cold but essentially normal objects such as planets and cold white-dwarf stars (so-called 'black dwarfs') has been largely discounted. Most cosmologists now agree that it consists of unknown, elusive subatomic particles nicknamed WIMPs (weakly interacting massive particles), more likely to be discovered in a particle physics laboratory than through a telescope.

A computer simulation maps the web of dark matter surrounding visible galaxies in the Universe.

Dark energy

In the late 1990s, cosmology was shaken by a discovery whose repercussions are still being felt. Astronomers using Type Ia supernovae (see page 338) to estimate the distance of remote galaxies found that they were fainter, and therefore more distant, than their red shift alone would predict (see page 348). They had previously expected these supernovae would be brighter than anticipated, and had planned to use this change to measure the rate at which the Universe is slowing down from the Big Bang. The discovery indicates that the reverse is happening – the growth of the Universe is speeding up, driven by an unknown force that has since been named 'dark energy'.

The nature of dark energy is still controversial – it may be a fifth fundamental force, or an aspect of spacetime itself. Further measurements have shown that dark energy has grown more powerful over time, and that cosmic expansion was indeed slowing down until around 6 billion years ago. Today, it accounts for some 68.3 per cent of energy in the Universe.

What is the Universe made of?

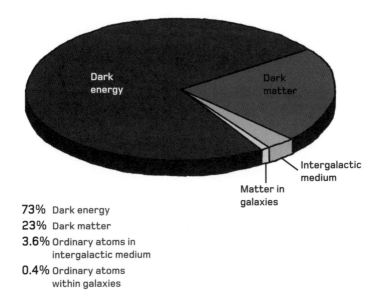

73% Dark energy
23% Dark matter
3.6% Ordinary atoms in intergalactic medium
0.4% Ordinary atoms within galaxies

Fates of the Universe

What is the ultimate destiny of the cosmos? For a long time, it seemed to be on a knife edge, with forces of expansion delicately balanced against inward gravitational pull. If gravity won out, we might see an eventual reversal of cosmic expansion, with the Universe collapsing back to a 'Big Crunch'. If not, then we were faced with either eternal expansion, or a gradual grinding to a halt – both scenarios in which the Universe would grow ever emptier and colder as the stars burnt out over many generations, and matter itself decayed back into its elementary particles. The discovery of dark energy (see page 402) seems to have tipped the scales in favour of eternal expansion – or possibly (if it continues to grow in strength), an eventual 'Big Rip' as the expansion of space becomes powerful enough to tear atoms themselves apart. However, new theories of cosmic creation, such as the idea that Big Bangs are triggered cyclically by collisions between higher-dimensional structures called 'branes', may offer ways for another Universe to rise from the ashes.

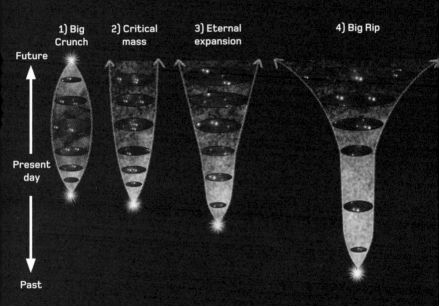

These schematics show four possible fates of the Universe: (1) collapse to a Big Crunch under its own gravitational attraction; (2) steady slowdown if the amount of mass is at a critical level; (3) eternal expansion thanks to weak gravity or dark energy; and (4) a dark-energy-driven 'Big Rip'.

The anthropic principle

Depending upon one's point of view, the existence of intelligent life in the Universe is both inevitable and hugely improbable. Given what we know about conditions across the cosmos – the values of fundamental forces and natural constants that allow atoms and molecules to bind in certain ways, influencing everything from the lifetimes of stars to the boiling and freezing points of water – we seem to live in a Universe ideally suited for life. It would be remarkable if ours was the only planet on which it had taken hold. Yet the odds of those forces and constants having taken on those precise values in the first place, when slight variations in any one of them might have created a Universe inimical to life, seem impossibly slender.

The 'anthropic principle', an approach to cosmology first put forward by Australian physicist Brandon Carter in 1973, attempts to address these issues. As its name suggests, it suggests that any cosmological model of the Universe must take account of the fact that it contains intelligent life. The simplest interpretation of this idea, called the 'weak anthropic principle', simply points out that we should not be too surprised at apparent 'fine-tuning' in nature, since if the Universe was not suitable for life, we would not be here to observe it.

In contrast, the 'strong anthropic principle' suggests that for one reason or another, the Universe *has* to be fine-tuned for life. Strong anthropic approaches to cosmology range from 'new physics' theories in which some as-yet-unknown physical laws control and guide the values of the observed constants, to 'design' theories in which the constants have been deliberately tuned by some outside intelligence – either alien or supernatural. In between are 'multiverse' theories that suggest all possible constants are able to play out in an infinite multitude of parallel Universes. Perhaps strangest of all, though, are the 'participatory' theories: rooted in some interpretations of quantum mechanics, these suggest that conscious observers are not just a by-product, but a *requirement* for the laws of physics to work as they do. If this last theory has any truth to it, then the Universe simply couldn't exist without us, and our fascination for the cosmos could have deeper roots than anyone has previously suspected.

Glossary

Active galaxy
A galaxy that emits large amounts of energy from its central regions, generated as matter falls into a supermassive black hole at the heart of the galaxy.

Asteroid
One of the countless rocky worlds of the inner solar system, largely confined in the main Asteroid Belt between the orbits of Mars and Jupiter.

Astronomical unit
A unit of astronomical measurement equivalent to Earth's average distance from the Sun – about 150 million km (93 million miles).

Binary star
A pair of stars in orbit around one another. Because such stars were usually born at the same time, they allow a direct comparison of the way that stars with different properties evolve.

Black hole
A superdense point in space, usually formed by a collapsing stellar core more than five times the mass of the Sun. A black hole's gravity is so powerful that even light cannot escape from it.

Celestial sphere
An imaginary spherical shell around the Earth that provides a useful 'model' for mapping the sky. The celestial equator and celestial poles are defined on the sphere in relation to Earth's own equator and poles.

Comet
A small chunk of rock and ice that typically orbits in the outer solar system. When comets are disturbed, they often fall into elliptical orbits that bring them closer to the Sun.

Constellation
Technically, one of 88

precisely defined areas of the sky. Traditionally, however, constellations are patterns in the sky produced by joining certain stars with imaginary lines.

Electromagnetic (e–m) radiation
A form of energy that consists of combined electric and magnetic waves, and is able to propagate itself across a vacuum at the speed of light.

Galaxy
An independent system of stars, gas and other material with a size measured in thousands of light years, containing anything from millions to billions of stars.

Globular cluster
A dense ball of ancient, long-lived stars, typically found in orbit around a galaxy such as the Milky Way.

Infrared
E-m radiation with slightly less energy than visible light. Infrared radiation is typically emitted by objects too cool to glow visibly.

Kuiper Belt
A doughnut-shaped ring of icy worlds orbiting beyond Neptune.

Light year
A unit of astronomical measurement equal to the distance travelled by light in one year – roughly 9.5 million million km (5.9 trillion miles).

Luminosity
A measure of a star's energy output, usually in comparison with the Sun. A star's luminosity in visible light is not necessarily equivalent to its entire energy output.

Magnitude
Apparent magnitude is a measure of a star's brightness as seen from Earth, using a simple number – the lower the number, the brighter the object. Absolute magnitude is a system that estimates a star's magnitude as seen from a standard distance.

Main sequence
A term used to describe the longest phase in a star's life, during which

it is relatively stable, and shines by fusing hydrogen (the lightest element) into helium (the next lightest) at its core.

Multiple star

A system of two or more stars in orbit around a shared 'centre of gravity' (pairs of stars are also called binaries). Most of the stars in our galaxy are members of multiple systems rather than individuals like the Sun.

Nebula

Any cloud of gas or dust floating in space. Nebulae are the material from which stars are born, and into which they are scattered again at the end of their lives.

Neutron star

The collapsed core of a supermassive star, left behind by a supernova explosion. Many neutron stars initially behave as pulsars.

Nova

A binary star system prone to dramatic outbursts, in which a white dwarf is pulling material from a companion star.

Nuclear fusion

The joining-together of light atomic nuclei (the central cores of atoms) to make heavier ones at very high temperatures and pressures, releasing excess energy in the process.

Oort Cloud

A spherical shell of dormant comets, up to two light years across, surrounding the entire solar system.

Open cluster

A large group of bright young stars that have recently been born from the same star-forming nebula.
.

Orbit

The typically elliptical path that a smaller body follows around a more massive one under the influence of gravity. Circular orbits are just an unusual form of ellipse.

Planet

A world that follows its own orbit around the Sun, is massive enough to pull itself into a spherical shape, and which has cleared the space around it of other objects (apart from satellites).

Planetary nebula
An expanding cloud of glowing gas sloughed off from the outer layers of a dying red giant star as it transforms into a white dwarf stellar remnant.

Pulsar
A rapidly spinning neutron star (collapsed stellar core) with an intense magnetic field that channels its radiation out along two narrow beams that sweep across the sky like a cosmic lighthouse.

Radio waves
The lowest-energy form of electromagnetic radiation, with the longest wavelengths. Radio waves are emitted by cool gas clouds in space, but also by violent active galaxies and pulsars.

Red dwarf
A star with considerably less mass than the Sun - small, faint and with a low surface temperature.

Red giant
A star passing through a phase of its life where its luminosity has increased hugely, causing its outer layers to expand and its surface to cool.

Spectrum
The rainbow-like band of light created by passing light through a prism or similar device. The prism bends light by different amounts depending on its wavelength and colour, so the spectrum reveals the precise intensities of light at different wavelengths.

Supernova
An enormous stellar explosion marking the death of a massive star.

Variable star
A star that changes its apparent brightness over time. Variations may be small or large, and may happen over very short or very long periods. They may be caused by interactions between stars, or by pulsations in the size and luminosity of an individual star.

Index

Quercus Editions Ltd
55 Baker Street,
7th Floor, South Block
London
W1U 8EW

First published in 2015

A catalogue record of this book is
available from the British Library

UK and associated territories:
ISBN 978 1 84866 723 5

Printed and bound in China

10 9 8 7 6 5 4 3 2 1

Picture credits 2:NASA, ESA and AURA/Caltech; 11: MarcelClemens/
Shutterstock; 23: NASA/JPL-Caltech/MSSS; 25: G.Gillet/ESO; 29:
Manamana/Shutterstock; 31: NASA/JPL-Caltech/K. Gordon (University
of Arizona); 33: NASA/CXC/SAO; 39: NASA; 41: NASA/Johns Hopkins
University Applied Physics Laboratory/Carnegie Institution of
Washington; 43: NASA/JPL; 45: Piotr Gatlik/Shutterstock; 49: NASA/
JPL/USGS; 51: NASA; 53: NASA/NEAR Project (JHU/APL); 55: NASA;
57: NASA/JPL-Caltech/University of Arizona; 65: NASA/JPL-Caltech/
UCAL/MPS/DLR/IDA; 67: NASA/JPL/Space Science Institute; 69: NASA/
JPL/University of Arizona; 71: NASA/JPL/University of Arizona; 73:
NASA/JPL/University of Arizona; 75: NASA/JPL; 77: NASA/JPL/DLR; 81:
NASA/JPL-Caltech/Space Science Institute; 82-3: NASA/JPL/Space
Science Institute; 85: NASA/JPL; 87: NASA/JPL-Caltech/Space Science
Institute; 89: NASA/JPL/Space Science Institute; 91: NASA/JPL/Space
Science Institute; 93: NASA/JPL/Space Science Institute; 95: NASA/
JPL/Space Science Institute; 97: ESA/NASA/JPL/University of Arizona;
99: NASA/JPL/Space Science Institute; 101: NASA/JPL/Space Science
Institute; 103: NASA/JPL/Space Science Institute; 105: NASA and ESA;
107: Lawrence Sromovsky, (Univ. Wisconsin-Madison), Keck Observatory;
109: NASA/JPL; 111: NASA/JPL-Caltech; 113: NASA/JPL; 115: NASA/
JPL; 117: NASA/JPL/USGS; 119: NASA/JPL; 123: NASA; 124: NASA, ESA,
and A. Schaller (for STScI); 276: T.A.Rector (NOAO/AURA/NSF) and
Hubble Heritage Team (STScI/AURA/NASA); 278: NASA, ESA, STScI, J.
Hester and P. Scowen (Arizona State University); 292: NASA, ESA, and G.
Bacon (STScI); 298: Daniel Price/Science Photo Library; 304: NASA, ESA
and H.E. Bond (STScI); 306: NASA, ESA and AURA/Caltech; 308: NASA,
The Hubble Heritage Team, STScI, AURA; 316: NASA, ESA, C.R. O'Dell
(Vanderbilt University), M. Meixner and P. McCullough (STScI); 318: NASA,
H.E. Bond and E. Nelan (Space Telescope Science Institute, Baltimore,
Md.); M. Barstow and M. Burleigh (University of Leicester, U.K.); and J.B.
Holberg (University of Arizona); 322: ESO; 326: NASA, ESA, J. Hester
and A. Loll (Arizona State University); 328: NASA/CXC/SAO/F.Seward
et al; 334: ESA, NASA and Felix Mirabel; 338: NASA/ESA, The Hubble Key
Project Team and The High-Z Supernova Search Team; 346: ESO/MPE/
Marc Schartmann; 350: D. Nidever et al., NRAO/AUI/NSF and A. Mellinger,
Leiden-Argentine-Bonn (LAB) Survey, Parkes Observatory, Westerbork
Observatory, and Arecibo Observatory; 352: Bill Schoening, Vanessa
Harvey/REU program/NOAO/AURA/NSF; 354: ESO; 358: NASA, ESA,
Hubble Heritage (STScI/AURA) et al.; 360: NASA/ESA and The Hubble
Heritage Team; 362: ESO/A. Milani; 364: NASA, ESA, and The Hubble
Heritage Team; 366: ESO/
WFI (Optical); MPIfR/ESO/APEX/A.Weiss et al. (Submillimetre); NASA/
CXC/CfA/R.Kraft et al. (X-ray); 370: NASA, ESA, and J. Lotz, M. Mountain,
A. Koekemoer, and the HFF team (STScI); 382: NASA / WMAP Science
Team; 390: Victor de Schwanberg/Science Photo Library; 398: Adolf
Schaller for STScI; 400: Springel et al. (2005).
All other illustrations by Tim Brown